Advances in Civil Engineering - Sustainable Materials and Resilient Structures

Edited by Ali Bahadori-Jahromi

Published in London, United Kingdom

Advances in Civil Engineering - Sustainable Materials and Resilient Structures
http://dx.doi.org/10.5772/intechopen.1003537
Edited by Ali Bahadori-Jahromi

Contributors
Abanoub M. AbdElmaseih, Ahmed Ashor Al-Saadi, Ali Bahadori Jahromi, Angel Antonio Ruiz Pico, Bamidele Charles Olaiya, Fariborz Nateghi-A, Ibtisam R. Karim, Mohsen Vatandoost, Mosleh Tohidi, Mustapha Muhammad Lawan, Nariman Hisham Halalo, Noboru Yuasa, Olaolu George Fadugba, Riyadh Alsultani, Ruth Noemi Salvador Reyes, Saleh I. Khassaf, Sol Estrella Espinal Bonilla

First published in London, United Kingdom, 2025 by IntechOpen
IntechOpen is the global imprint of INTECHOPEN LIMITED, registered in England and Wales, registration number: 11086078, 167-169 Great Portland Street, London, W1W 5PF, United Kingdom

For EU product safety concerns: IN TECH d.o.o., Prolaz Marije Krucifikse Kozulić 3, 51000 Rijeka, Croatia, info@intechopen.com or visit our website at intechopen.com.

British Library Cataloguing-in-Publication Data
A catalogue record for this book is available from the British Library

Advances in Civil Engineering - Sustainable Materials and Resilient Structures
Edited by Ali Bahadori-Jahromi
p. cm.

This title is part of the Civil Engineering Book Series, Volume 12
Topic: Construction Engineering
Series Editor: Assed Haddad
Topic Editor: Amin Akhnoukh

Print ISBN 978-1-83769-875-2
Online ISBN 978-1-83769-874-5
eBook (PDF) ISBN 978-1-83769-876-9
ISSN 3029-0287

If disposing of this product, please recycle the paper responsibly.

IntechOpen Book Series

Civil Engineering

Volume 12

Aims and Scope of the Series

Civil engineering is a traditional field of engineering from which most other branches of engineering have evolved. It comprises traditional sub-areas like transportation, structures, construction, geotechnics, water resources, and building materials. It also encompasses sustainability, risk, environment, and other concepts at its core. Historically, developments in civil engineering included traditional aspects of architecture and urban planning as well as practical applications from the construction industry. Most recently, many elements evolved from other fields of knowledge and topics like simulation, optimization, and decision science have been researched and applied to increase and evolve concepts and applications in this field. Civil engineering has evolved in the last years due to the demands of society in terms of the quality of its products, modern applications, official requirements, and cost and schedule restrictions. This series addresses real-life problems and applications of civil engineering and presents recent, cutting-edge research as well as traditional knowledge along with real-world examples of developments in the field.

Meet the Series Editor

Professor Assed N. Haddad is a Civil Engineer with a degree from the Federal University of Rio de Janeiro (UFRJ) earned in 1986, as well as a Juris Doctor degree from the Fluminense University Center earned in 1993, and a Master's degree in Civil Engineering from the Fluminense Federal University (UFF) obtained in 1992. He completed his Ph.D. in Production Engineering from COPPE / Federal University of Rio de Janeiro in 1996. Professor Haddad's academic pursuits have taken him to postdoctoral stays at the University of Florida, USA in 2006; at the Universitat Politècnica de Catalunya, Spain in 2010; and at the University of New South Wales Sydney, Australia in 2019. Currently, he serves as a Full Professor at the Federal University of Rio de Janeiro. He has held visiting professorships at various institutions including the University of Florida, Universitat Politècnica de Catalunya, Universitat Rovira i Virgili, and Western Sydney University. His research expertise encompasses Civil, Environmental, and Production Engineering, with a primary focus on the following topics: Construction Engineering and Management, Risk Management, and Life Cycle Assessment. He has been the recipient of research grants from the State of Rio de Janeiro, Brazil: CNE FAPERJ from 2019 to 2022 and from 2023 to 2025. Additionally, his research grants obtained from the Brazilian Government CNP since 2012 last to this date. Professor Haddad has been involved in several academic endeavors, being the Guest Editor of the International Journal of Construction Management; MDPI's Sustainability, Energies, and Infrastructures; Associate Editor at Frontiers in Built Environment / Sustainable Design and Construction; Guest Editor at Frontiers in Built Environment / Construction Management; and Academic Editor of the Journal of Engineering, Civil Engineering Section of Hindawi. He is currently a Professor of the Environmental Engineering Program at UFRJ and the Civil Engineering Program at UFF.

Meet the Volume Editor

Professor Ali Bahadori-Jahromi is an experienced academic and professional in Civil Engineering, currently serving as a Professor at the University of West London. He leads the Building Performance and Climate Change research group. He oversees Ph.D. programs in engineering, focusing on advancing knowledge in sustainable construction, energy efficiency, and structural analysis. As a Chartered Engineer (CEng) and Fellow of the Institution of Civil Engineers (FICE), Professor Bahadori-Jahromi has developed accredited civil engineering courses and promoted climate-resilient building practices. His work includes publications and patents that contribute to both academia and industry. He is committed to mentoring future engineers and researchers, supporting their efforts to address challenges in the built environment.

Contents

Preface

The evolution of construction materials and technologies has shaped the modern built environment. Among these advancements, reinforced concrete stands as a cornerstone, valued for its exceptional strength, durability, and versatility. This edited volume, *Advances in Civil Engineering – Sustainable Materials and Resilient Structures*, presents contemporary research and developments addressing the complex challenges faced by practitioners, researchers, and policymakers in this field. As the editor, it has been an honour to work with distinguished experts whose contributions showcase various innovative solutions and perspectives.

This volume is structured to provide a comprehensive understanding of advancements in civil engineering materials and technologies, spanning foundational principles to pioneering innovations. The chapters cover diverse yet interconnected topics, collectively advancing construction practices.

Chapter 1: "Analysis of Asphalt Mixtures Using Construction and Demolition Wastes as Filler"

This chapter examines the incorporation of recycled materials into asphalt mixtures. It highlights the potential of construction and demolition wastes as sustainable fillers, enhancing asphalt's mechanical properties and resistance while addressing environmental concerns.

Chapter 2: "Building Information Modeling (BIM) Implementation and Practices in Construction Industry: A Review"

This chapter reviews the global adoption of BIM technologies, exploring their transformative impact on communication, efficiency, and integration within the construction industry. It offers insights into emerging trends and practices.

Chapter 3: "Dynamic Response Analysis of Coastal Bridge Members Exposed to Water Forces and Earthquakes"

This chapter investigates the behaviour of coastal bridges under dynamic forces. Using finite element analysis, it evaluates the interplay of hydrodynamic forces, wave currents, and seismic activity on the stability of pile foundations.

Chapter 4: "Effects of Backfill Types on the Behavior of Buried Concrete Sewerage Pipelines in Soil"

This chapter focuses on the influence of backfill materials on pipeline infrastructure. Advanced modelling techniques assess soil-pipeline interaction, revealing how compaction and material properties impact pipeline stability.

Chapter 5: "Nanohybrid Polymer Concrete: Advancing towards Sustainable and Durable Infrastructure"

This chapter explores the potential of nanotechnology in enhancing polymer concrete. Integrating unsaturated polyester resin and nanomaterials demonstrates advancements in mechanical properties, paving the way for sustainable and durable construction solutions.

Chapter 6: "Perspective Chapter: Comprehensive and New Approximate Analysis and Design Techniques for Reinforced Concrete Structural Elements"

This chapter introduces innovative analysis and design methodologies for reinforced concrete systems. It focuses on new floor systems, high-rise buildings, and tubular and outrigger systems, offering practical tools for structural optimization.

Chapter 7: "Seismic Retrofitting Reinforced Concrete Structures with Precast Pre-Stressed Concrete Braces: An Overview"

This chapter addresses the critical challenge of seismic resilience. It presents techniques for retrofitting structures using precast and pre-stressed concrete braces, supported by finite element modelling, to improve structural safety and strength.

Chapter 8: "Surface Layer Concrete"

This chapter examines the deterioration and inhomogeneity of concrete surfaces. It discusses the implications of surface properties on structural integrity, providing strategies for mitigation and extending the lifespan of concrete structures.

This volume represents the collective effort of a dedicated team of authors who have shared their expertise and innovative research. I am deeply grateful to the team at IntechOpen for their guidance and support throughout this process.

Finally, I extend my gratitude to my colleagues, collaborators, and students, whose thought-provoking discussions have inspired me. This book is dedicated to the global community of engineers, researchers, and professionals committed to advancing sustainable materials and resilient structures, ensuring a sustainable and resilient future for future generations.

Ali Bahadori-Jahromi
Professor of Civil Engineering,
University of West London (UWL),
London, UK

Fellow of the Institution of Civil Engineers (FICE),
Chartered Engineer (CEng),
Member of the Joint Board of Moderators (JBM),
London, UK

Chapter 1

Analysis of Asphalt Mixtures Using Construction and Demolition Wastes as Filler

Angel Antonio Ruiz Pico, Ruth Noemi Salvador Reyes and Sol Estrella Espinal Bonilla

Abstract

The objective of this research was to evaluate the physical and mechanical properties of asphalt mixtures using construction and demolition waste (CDW) as fillers for flexible pavement projects designed for light traffic. Asphalt briquettes were produced, incorporating CDW filler, and their performance was assessed in comparison with standard mixtures. The study identified the properties of CDW necessary for its use as filler, determining the optimal percentages to be incorporated into the asphalt mixtures. Mix designs were developed for both the standard sample and for mixtures containing 1, 3, 5, and 7% CDW filler. The results showed that while 1% CDW closely mirrors the properties of the standard asphalt mix and complies with the "General Technical Specifications for Construction" (2013), the optimal performance was achieved at 5% CDW, which exhibited the highest resistance and stability under load. However, not all filler percentages met the required properties for asphalt mixtures, highlighting the importance of proper dosage for effective performance. The findings suggest that incorporating CDW as filler in asphalt mixes can offer a sustainable alternative for flexible pavement construction, provided that filler proportions are carefully controlled.

Keywords: asphalt mixtures, CDW, filler, construction, demolition wastes

1. Introduction

Currently, the construction sector generates a large amount of waste due to the different projects and civil works that have been carried out over the years, as well as the termination of the useful life of old constructions that produce a large amount of Construction and Demolition Waste [1]. The increase in construction and demolition waste allows mechanisms to reuse these materials in the generation of new products with applications in civil engineering [2].

The reuse of CDW is one of the main issues in the construction industry [3] and is generated in the city for the production of recycled materials, being a key alternative for the sustainability of the construction industry. The proper use of these can contribute positively to the reduction of environmental impact, avoiding the depletion of raw materials, reducing the extraction of these materials from quarries, and reducing the volume of waste that ends up in landfills, achieving the care of the environment [4].

Globally, it is estimated that more than 25 billion tons per year of concrete are produced worldwide, which is equivalent to an average of 1 m3 per person, and 10 billion tons per year of CDW are produced worldwide, of which the United States contributes approximately 700 million tons, China contributes about 2.3 billion tons, and the European Union more than 800 million tons [5].

According to a study published in 2015, the annual global weight of landfilled CDW currently amounts to 1.2 billion tons [6], at the end of a project, it is estimated that the production of construction waste is between 35% and 40% of total waste, which, at the end of the project, ends up in illegal dumps or final disposal sites without being properly used [7].

Building materials are difficult to dispose of because they may contain hazardous materials, such as asbestos, heavy metals, persistent organic substances, among others [8]. However, its implementation can be favorable for the improvement of management processes [2] because CDW materials can be implemented in raw materials for the construction of road and civil projects.

Road projects have become more important in the sector, due to the need of people to travel to different cities and/or countries, that is why pavements are constantly exposed to traffic loads and environmental factors, causing deterioration and therefore a decrease in their quality, having an increase in maintenance costs. Therefore, it is essential to preserve the quality of roads and pavements. In the case of Colombia, 204,855 km of roads are reported, of which 17,434 km are distributed in the primary network and 187,421 km in the secondary and tertiary network. In the case of the paved road network, 47.2% is in very good to good condition, 34.3% in fair condition, and 18.5% in poor to very poor condition [9].

Correct estimation of construction and demolition waste enables the establishment of an effective management system at both the project and regional levels [10]. In general, there is a lack of knowledge of the process of generation and estimation of CDW, in order to achieve efficient recycling management.

In Peru, in general, there are no systematized studies on the reuse of CDW as filler, the support of the state is important to promote on-site recycling, although it is true that it is difficult to have equipment that facilitates the management of CDW.

The objective of this research is to show, through a practical example carried out in the laboratory with its respective tests, the evaluation of the physical and mechanical properties of asphalt mixtures using CDW as filler for flexible pavement projects for light traffic, through concrete blocks coming from dumps belonging to concrete for demolition false floors.

2. Material and methods

2.1 Selection of materials

Asphalt mixtures: Asphalt mixtures consist mainly of aggregates and asphalt. It is possible to incorporate rubber powder from used tires into these materials without compromising their structural integrity or capacity [11], aggregates represent over

90% of the total mass of asphalt mixtures, with a production rate of approximately 435,106 tons per year in the European Union [12]. The primary source of these aggregates is quarrying, which has significant environmental, economic, and energy costs [13].

Approximately 70% of the asphalt, by weight, is used in road construction due to its waterproofing and binding properties. However, asphalt is susceptible to high temperatures, necessitating modification prior to its application in construction and infrastructure projects [14]. One such modification involves the incorporation of crumb rubber from recycled tires, which can improve the performance and mechanical properties of the asphalt mixture in road infrastructure projects like pavements [8]. This results in a surface treatment that is cost-effective, efficient, and environmentally friendly [3].

Construction and demolition waste (CDW): Construction and demolition waste (CDW) refers to materials generated during the construction, remodeling, repair, or demolition of buildings, bridges, pavements, and other structures. According to the Scottish Environment Protection Agency (2014), these materials also encompass the use of energy, materials, and labor that do not add value to the construction process [9]. CDW comprises a variety of construction materials, including concrete (approximately 40%) and ceramics (about 30%) [5]. A significant portion (65–75%) of these wastes originates from stone-based materials, such as concrete, mortars, bricks, tiles, and flagstones [15], making them suitable for recycling in certain cases [16].

Internationally, CDW can be classified according to its source [17]:

- Excavation materials: soil, sand, gravel, rocks, etc.
- Construction and maintenance of civil works: asphalt, sand, gravel, metals, etc.
- Demolition materials: concrete blocks, bricks, plaster, porcelain, and lime plaster.

Research shows that incorporating 20% CDW in asphalt mixtures can yield satisfactory technical performance for paving urban streets. The performance of these mixtures is influenced by the properties, composition, and proportions of CDW aggregates [18]. Furthermore, most studies focus on reusing construction waste for concrete production or as raw materials for cement [3].

Filler: Fillers are commercial products of fine particulate nature, such as cement or fly ash from thermal power plants, or limestone powder, commonly used in asphalt mixtures. They play a crucial role in determining the mechanical properties and behavior of the asphalt mixture, as they contribute to internal friction between particles and facilitate compaction, thus acting as a binder between larger aggregates.

Several studies have explored the use of different fillers from CDW and solid waste as substitutes for mineral fillers in concrete mixtures. These studies suggest that such fillers do not negatively affect the properties of concrete, and in some cases, they enhance properties such as waterproofing and densification [1].

2.2 CDW sample preparation

The study analyzed a sample of 75 asphalt mixture briquettes, prepared according to the "Roads Manual" (EG-2013), which aligns closely with the European standard (Road Law - Law 37/2015). The number of blows applied to the samples was determined based on the average traffic previously estimated.

2.3 Process for obtaining CDW

To obtain the CDW material, critical collection points were identified, mapped, and visited in various dumping sites in Lambayeque (Peru) (**Figure 1**).

The materials selected came from key points, particularly focusing on concrete blocks. It was decided to use concrete from demolished false floors, as the literature indicates that such materials are high-quality waste with minimal impurities. The selected material was subsequently reduced in size using a crusher, followed by passing it through the Los Angeles abrasion machine to achieve the desired granularity, ensuring it passed through mesh No. 50. This material was used as a replacement for the filler (**Figure 2**).

The asphalt mixture design was based on the Marshall methodology, which involves determining the percentage of stone materials and filler to be embedded in asphalt cement (PEN), which must be heated to a temperature range of 150–180°C. The asphalt mixture designs for samples containing 1, 3, 5, and 7% CDW were formulated and tested.

Figure 1.
Collection point for CDW.

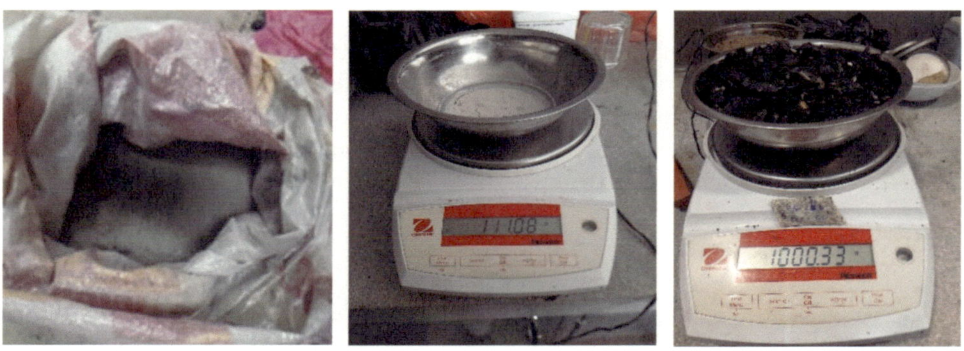

Figure 2.
Crushed concrete waste material.

In the preparation of the standard asphalt mixture, the proportions of coarse aggregate (60%), fine aggregate (39%), and filler (1%) were maintained constant. Different percentages of asphalt cement (4.0, 4.50, 5.0, 5.50, and 6.0%) were incorporated to study their effects.

Tables 1–4 present the results for the standard asphalt mixture incorporating 1, 5, and 7% CDW, along with a detailed characterization of the samples based on varying asphalt cement percentages.

2.4 Tests performed

The following tests are performed according to the standard specifications (**Table 5**).

Mezcla Asfaltica Para 1% RCD			
Cemento Asfáltico	**Agregado Grueso**	**Agregado Fino**	**Filler (RCD)**
4.0%	60%	39%	1%
4.5%	60%	39%	1%
5.0%	60%	39%	1%
5.5%	60%	39%	1%
6.0%	60%	39%	1%

Table 1.
Design of asphalt mix for 1% RCD.

Mezcla Asfaltica Para 3% RCD			
Cemento Asfáltico	**Agregado Grueso**	**Agregado Fino**	**Filler (RCD)**
4.0%	58%	39%	3%
4.5%	58%	39%	3%
5.0%	58%	39%	3%
5.5%	58%	39%	3%
6.0%	58%	39%	3%

Table 2.
Design of asphalt mix for 3% RCD.

Mezcla Asfaltica Para 5% RCD			
Cemento Asfáltico	**Agregado Grueso**	**Agregado Fino**	**Filler (RCD)**
4.0%	57%	38%	5%
4.5%	57%	38%	5%
5.0%	57%	38%	5%
5.5%	57%	38%	5%
6.0%	57%	38%	5%

Table 3.
Design of asphalt mix for 5% RCD.

	Mezcla Asfaltica Para 7% RCD		
Cemento Asfáltico	**Agregado Grueso**	**Agregado Fino**	**Filler (RCD)**
4.0%	56%	37%	7%
4.5%	56%	37%	7%
5.0%	56%	37%	7%
5.5%	56%	37%	7%
6.0%	56%	37%	7%

Table 4.
Design of asphalt mix for 7% RCD.

Aggregate test	Asphalt mix testing
Granulometry (MTC E 107)	Specific Gravity (ASTM D 854)
Specific Gravity and Coarse Aggregate Absorption (MTC E 206)	Marshall Test (MTC E 504) Rice Test (ASTM D 2071 and AASHTO T 209)
Abrasion test (MTC E 20)	
Flat and elongated particles (MTC E 221)	
Soluble salts (MTC E 219)	
Durability test (ASTM C 88)	
Fractured faces (MTC E 210)	
Sand equivalent (ASTM D 2419)	
Plasticity index (MTC E 111)	

Table 5.
Standard specifications.

3. Results and discussion

In order to obtain the mix design for the standard sample and for the samples with the addition of 1, 3, 5, and 7% of CDW filler, the relative specific weight test for solids of the RDC was previously performed, where the value of the relative specific weight test for solids was 2.479 g/cm3, then the test for organic matter in soils was developed, where the laboratory determined that the value for organic matter for the CDW sample due to loss on ignition was 2.1%.

The corresponding tests were performed on the coarse aggregate according to the E.G. 2013 standard. In the specific gravity and absorption test, the aggregate complied with the maximum permitted value of 1%. For the abrasion test with the Los Angeles machine, the wear obtained was 16.4%, within the required limit of 40%. As for the determination of flat and elongated particles, the indices were 2.75% for elongation and 5.77% for flattening, both within the maximum of 10% established by the standard. The soluble salt content was 0.03%, well below the 0.5% limit. In the magnesium sulfate durability test, a value of 6% was obtained. Finally, the fractured face test showed that 81.2% of the particles presented one fractured face, and 83.7% presented two or more, complying with the normative parameters.

Subsequently, the tests corresponding to the fine aggregate were carried out comparing the results with the E.G. 2013 standard. In the sand equivalence test, a percentage of 61% was obtained, complying with the required minimum of 60%. The specific gravity test on fine aggregate yielded an absorption value of 0.19%, within the normative parameter that establishes a maximum of 0%. As for the content of soluble salts, the result was 0.02%, a value accepted by the standard. In the magnesium sulfate durability test, it was determined that the natural sand has a durability of 7%, meeting the regulatory requirements. Finally, the plasticity test showed that the yellow sand has no liquid or plastic limit, so its plasticity index is null, confirming that it is a non-plastic material.

Likewise, the design of the standard asphalt mix was elaborated, where certain elements were considered for its analysis: crushed gravel 3/4" at 60%, crushed sand 1/4" at 39%, and cement (Filler) at 1%.

After performing the corresponding tests mentioned above, the results of the standard mix design with 1% CDW, 3% CDW, 5% CDW, and 7% CDW will be obtained with their respective graphical representation of the asphalt design.

Table 6 shows the results of the master mix design with 1% hydraulic cement filler for all variations in percentages of asphalt cement, which are 4, 4.5, 5, 5.5, and 6%. The data obtained complies with the regulations, having more favorable results in the PEN variations of 5.5 and 6%.

Table 7 shows the results obtained in the mix design with the addition of 1% of concrete CDW filler for all variations in percentages of asphalt cement, which are 4, 4.5, 5, 5.5, and 6%. The data obtained complies with the regulations, having more favorable results in the PEN variations of 1 and 5.5%.

In **Figure 3**, the percentage of voids is 2.8, having a stability of 1800 kg and achieving an asphalt dust ratio of 0.98, the percentage being lower than **Figure 4**.

In **Table 8**, the results were obtained in the mix design with the addition of 3% of concrete CDW filler for all variations in percentages of asphalt cement, which are 4, 4.5, 5, 5.5, and 6%. The data obtained complies with the regulations, having more favorable results in the PEN variations of 6% for the stability/flow relationship and 4.5% for the other parameters.

In **Figure 5**, the percentage of voids is 2.6, having a stability of 2220 kg and achieving an asphalt powder ratio of 0.92.

In **Table 9**, the results were obtained in the mix design with the addition of 5% of concrete CDW filler for all variations in percentages of asphalt cement, which are 4, 4.5, 5, 5.5, and 6%. The data obtained complies with the regulations, having more favorable results in the PEN variations of 4.5 and 5.5%.

In **Figure 6**, the percentage of voids is 3, having a stability of 2300 kg and achieving an asphalt dust ratio of 0.89, a lower percentage than **Figure 5**.

In **Table 10**, the results were obtained in the mixture design with the addition of 3% of concrete CDW filler for all variations in percentages of asphalt cement, which are 4, 4.5, 5, 5.5, and 6%. The data obtained complies with the regulations, having more favorable results in the PEN variations of 4 and 4.5% for the flow and 5–6% for the other parameters.

In **Figure 7**, the percentage of voids is 3, having a stability of 2300 kg and achieving an asphalt dust ratio of 0.86, being the lowest value in the analyses.

The results obtained in the mix design with the addition of 3% concrete CDW filler for all variations in asphalt cement percentages are 4, 4.5, 5, 5.5, and 6%. The data obtained complies with the regulations, with the most favorable results in the PEN variations of 4 and 4.5% for the flow and 5–6% for the other parameters.

Results								
Relation Pen-Filler	**Unit weight (gr/cm^2)**	**Voids (%)**	**Mineral aggregate voids (%)**	**C.A filled voids (%)**	**Flow (mm)**	**Stability (Kg)**	**Asphalt dust ratio**	**Flow stability ratio**
100	**2.28**	**4.50**	**22.00**	**68.00**	**12.70**	**1755.68**	**1.23**	**1809.22**
1–4%	1.99	1.11	37.56	97.06	22.00	1748.33	1.23	3129.01
1–4.5%	2.13	2.65	56.99	95.36	25.67	1501.45	1.23	2297.23
1–5%	2.22	2.32	55.77	95.87	26.33	1671.02	1.23	2505.54
1–5.5%	2.21	1.42	56.01	97.47	26.33	1976.94	1.23	2995.24
1–6%	2.20	0.99	56.06	98.25	20.67	1849.86	1.23	3629.41

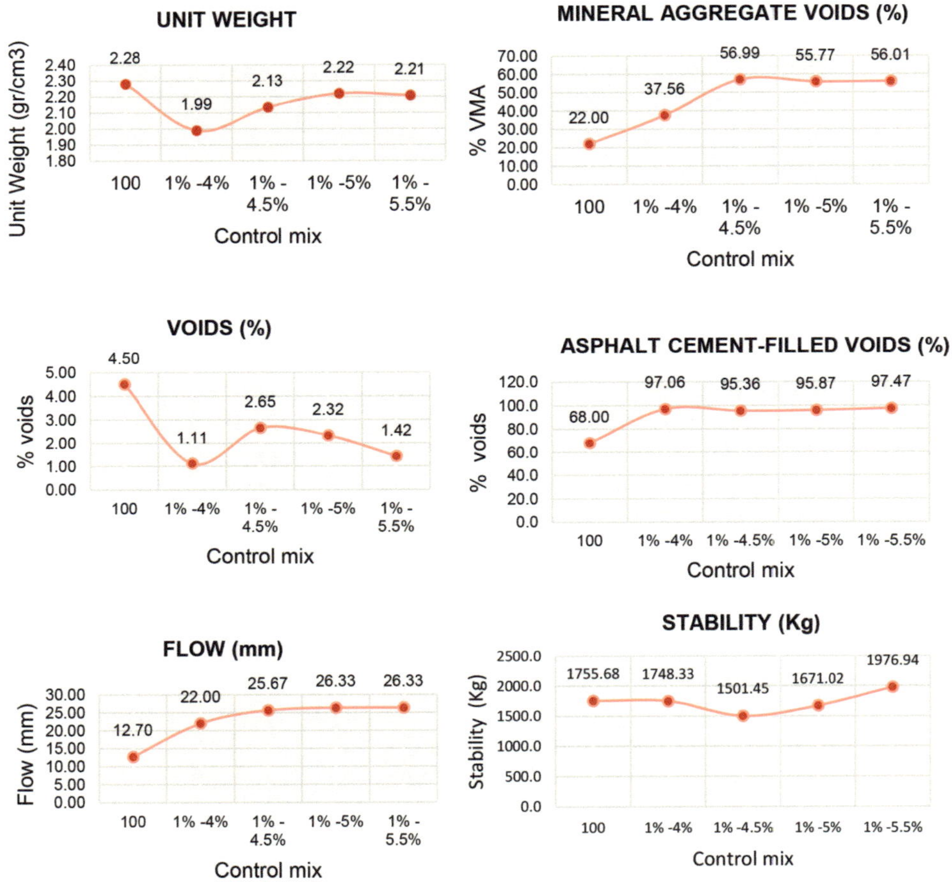

Table 6.
Analysis of the results of the pattern mix design.

A comparative table of the tests was made, obtaining results such as unit weight, void percentages, mineral aggregate voids, A.C. filled voids, flow, stability, and asphalt powder ratio (**Table 11**).

Results								
Relation Pen-Filler	**Unit weight (gr/cm^2)**	**Voids (%)**	**Mineral aggregate voids (%)**	**C.A filled voids (%)**	**Flow (mm)**	**Stability (Kg)**	**Asphalt dust ratio**	**Flow stability ratio**
100	**2.28**	**4.50**	**22.00**	**68.00**	**12.70**	**1755.68**	**1.23**	**1809.22**
1–4%	1.97	1.91	38.28	95.02	22.33	1821.33	1.23	3353.26
1–4.5%	2.20	2.33	55.57	95.80	26.33	1818.17	1.23	2833.13
1–5%	2.20	3.02	56.08	94.62	27.67	1913.99	1.23	2789.07
1–5.5%	2.20	2.44	56.06	95.66	23.33	1834.72	1.23	3090.70
1–6%	2.24	3.04	55.30	94.53	31.67	1691.53	1.23	2302.94

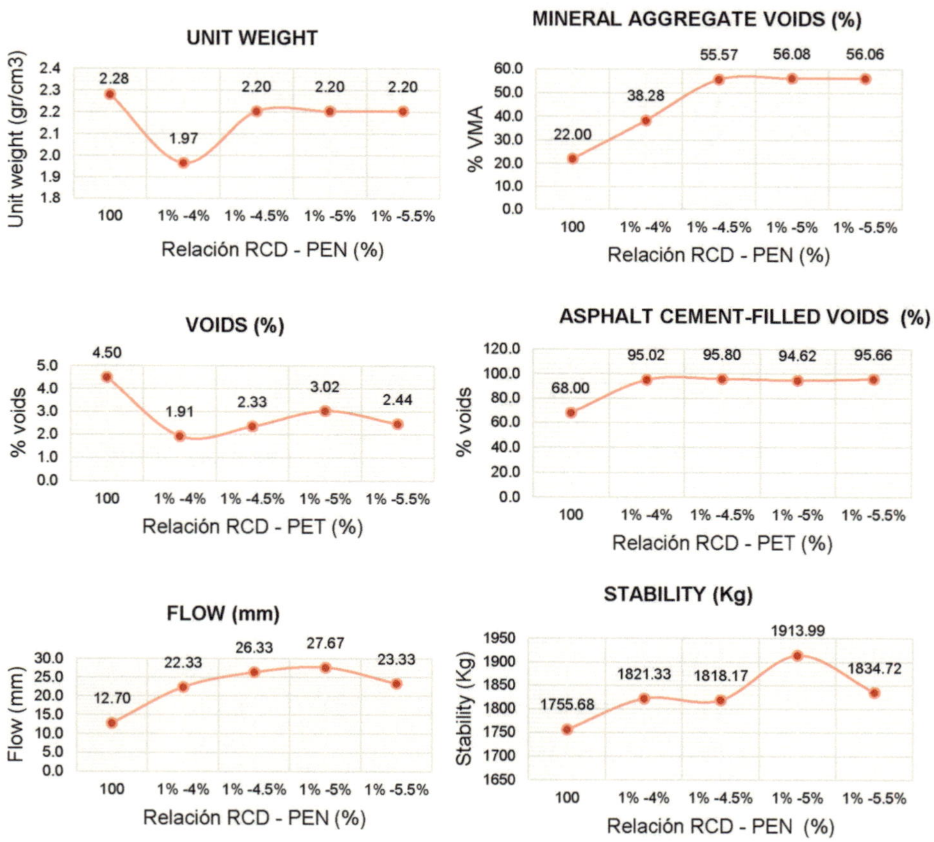

Table 7.
Analysis of mixture design results with 1%RCD - %PEN.

It can be seen that 1% CDW is the most similar to the standard asphalt mix and is within the parameters established in the EG - 2013 standard. However, the optimum percentage is 5% CDW because it shows results with greater resistance, as with the percentages obtained in [9] the percentages obtained favor the stability of the asphalt mixes.

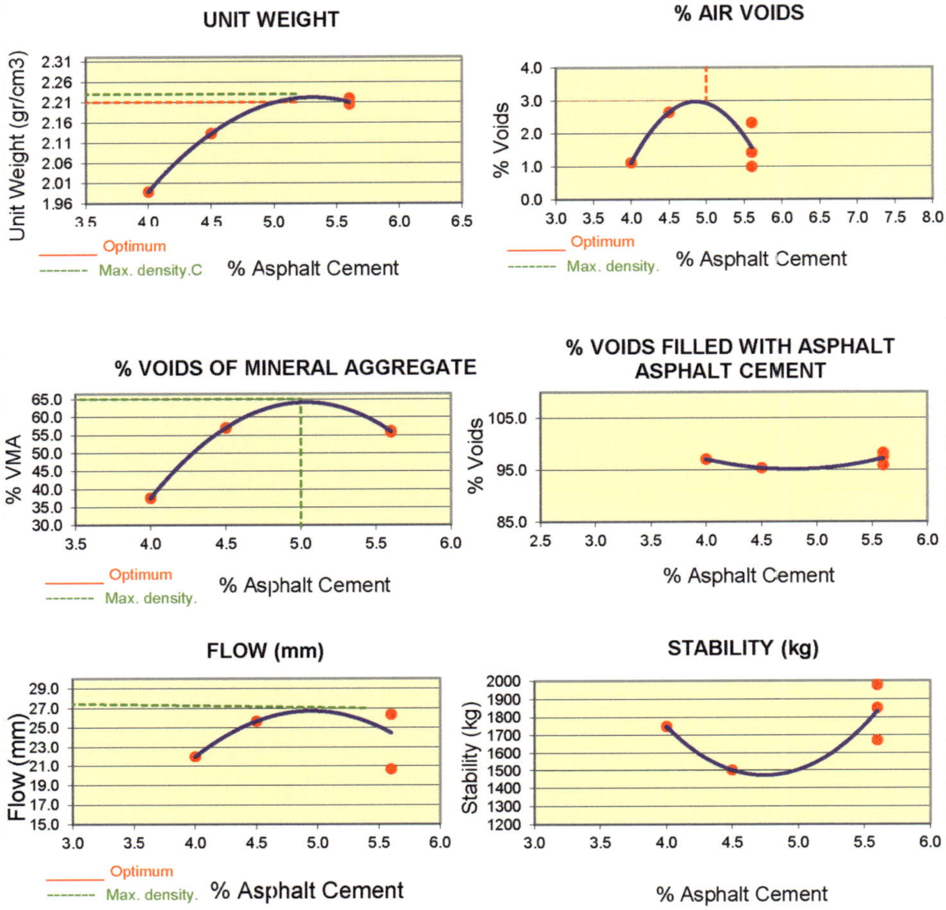

RESULTS	
Optimum Content C.A	5.27
Unit weight (gr/cm2)	2.23
Voids (%)	3.00
Mineral aggregate voids(%)	65.00
Voids filled with C.A (%)	95.00
Flow (mm)	7.11
Stability (Kg)	1500.00
Stiffness Index (kg/cm)	3250.00
Asphalt Dust Ratio	0.95

Figure 3.
Graphicalrepresentation of asphalt design sample pattern.

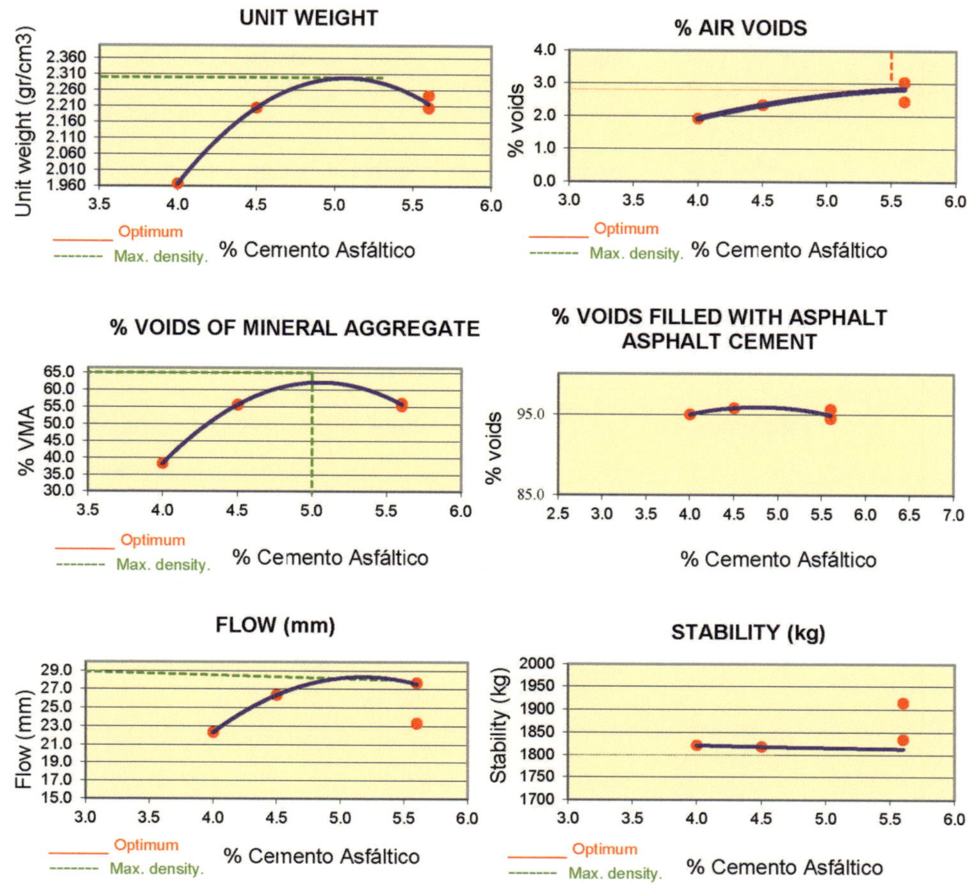

RESULTS	
Optimum Content C.A	5.43
Unit weight (gr/cm2)	2.30
Voids (%)	2.80
Mineral aggregate voids(%)	65.00
Voids filled with C.A (%)	95.00
Flow (mm)	7.62
Stability (Kg)	1800.00
Stiffness Index (kg/cm)	3250.00
Asphalt Dust Ratio	0.98

Figure 4.
Graphical representation of asphalt design with 1%RCD.

Results								
Relation Pen-Filler	**Unit weight (gr/cm^2)**	**Voids (%)**	**Mineral aggregate voids (%)**	**C.A filled voids (%)**	**Flow (mm)**	**Stability (Kg)**	**Asphalt dust ratio**	**Flow stability ratio**
100	**2.28**	**4.50**	**22.00**	**68.00**	**12.70**	**1755.68**	**1.23**	**1809.22**
1–4%	1.96	2.70	39.86	93.27	20.33	2235.00	1.23	4468.24
1–4.5%	2.21	2.62	55.47	95.30	26.67	1843.98	1.23	2741.73
1–5%	2.18	3.57	56.57	93.77	19.67	1950.90	1.23	4188.18
1–5.5%	2.08	2.72	58.53	95.36	30.00	1899.46	1.23	2598.59
1–6%	2.14	3.35	57.41	94.19	25.00	1951.12	1.23	3134.26

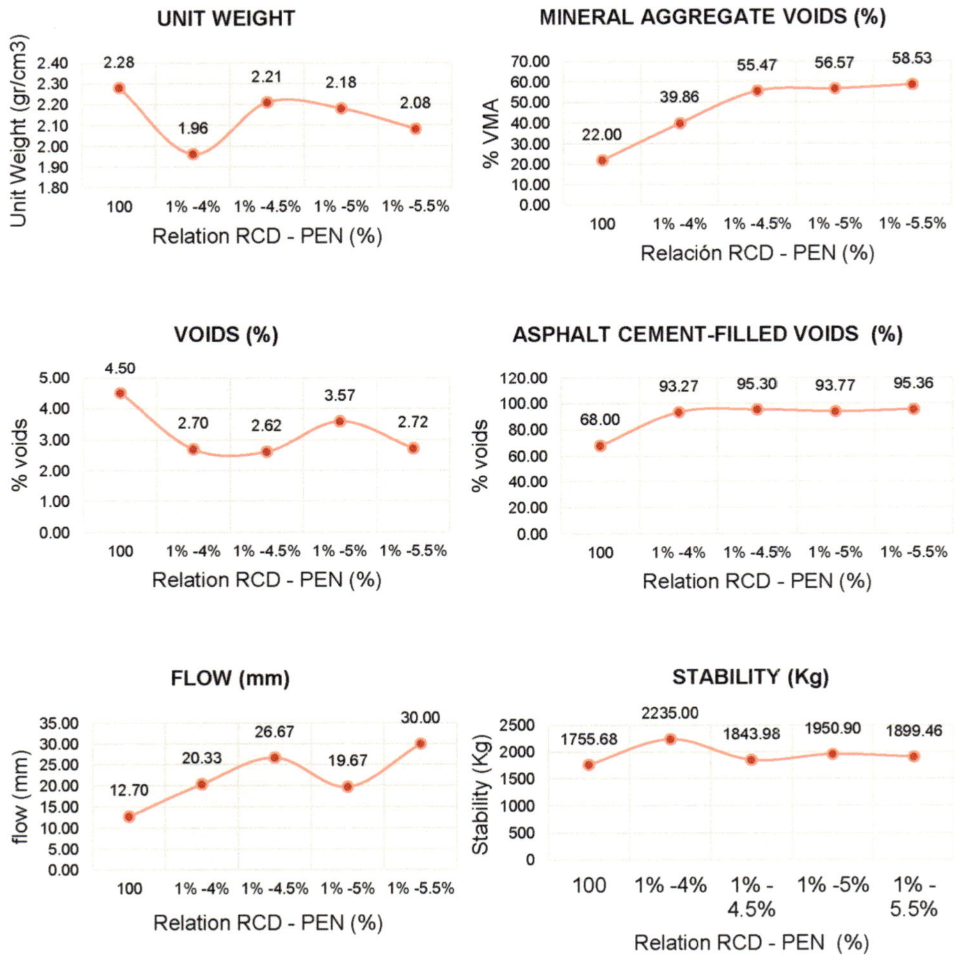

Table 8.
Analysis of the results of the mix design with 3%RCD - %PEN.

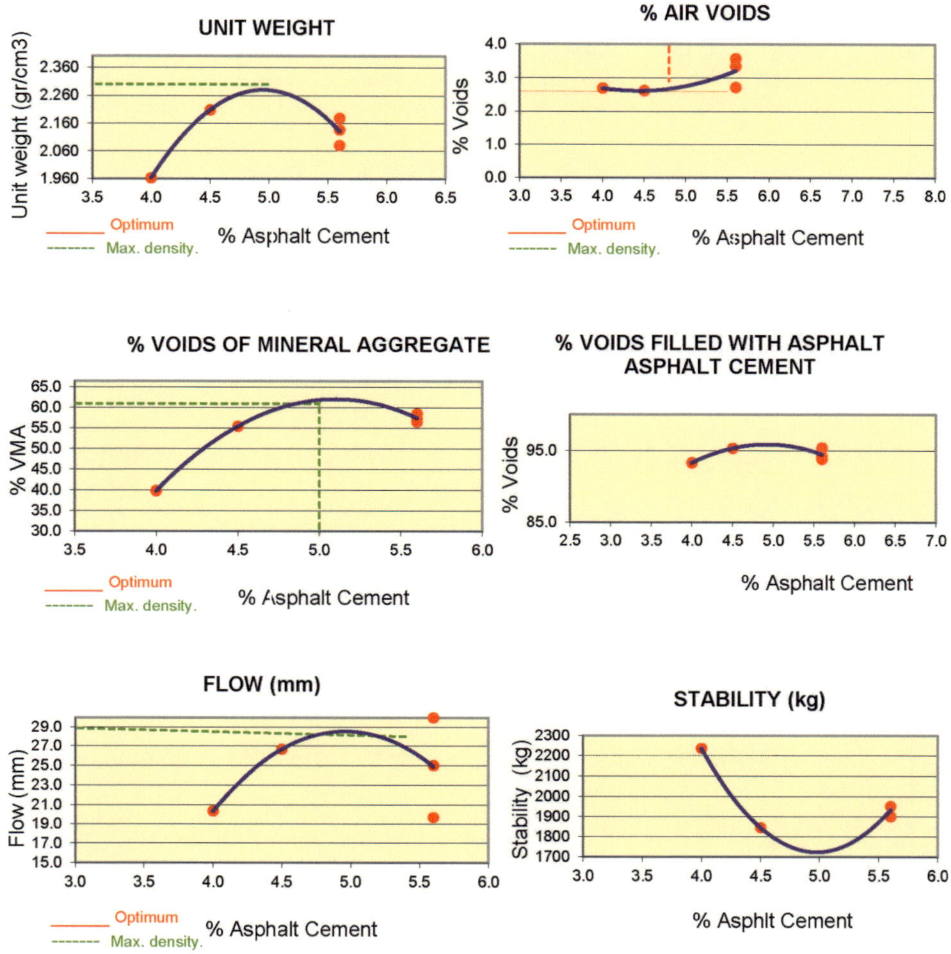

RESULTS	
Optimum Content C.A	5.13
Unit weight (gr/cm2)	2.30
Voids (%)	2.60
Mineral aggregate voids(%)	61.00
Voids filled with C.A (%)	95.00
Flow (mm)	7.62
Stability (Kg)	2200.00
Stiffness Index (kg/cm)	3250.00
Asphalt Dust Ratio	0.92

Figure 5.
Graphical representation of asphalt design with 3%RCD.

Results								
Relation Pen-Filler	**Unit weight (gr/cm^2)**	**Voids (%)**	**Mineral aggregate voids (%)**	**C.A filled voids (%)**	**Flow (mm)**	**Stability (Kg)**	**Asphalt dust ratio**	**Flow stability ratio**
100	**2.28**	**4.50**	**22.00**	**68.00**	**12.70**	**1755.68**	**1.23**	**1809.22**
1–4%	2.20	2.42	34.10	92.91	20.33	2310.00	1.23	4525.67
1–4.5%	2.21	2.82	55.36	94.91	21.67	1907.97	1.23	3673.96
1–5%	2.22	2.93	55.78	94.75	18.00	2005.84	1.23	4501.11
1–5.5%	2.22	3.02	55.82	94.59	18.67	1776.98	1.23	4005.51
1–6%	2.22	2.86	55.82	94.88	26.67	1435.09	1.23	2325.51

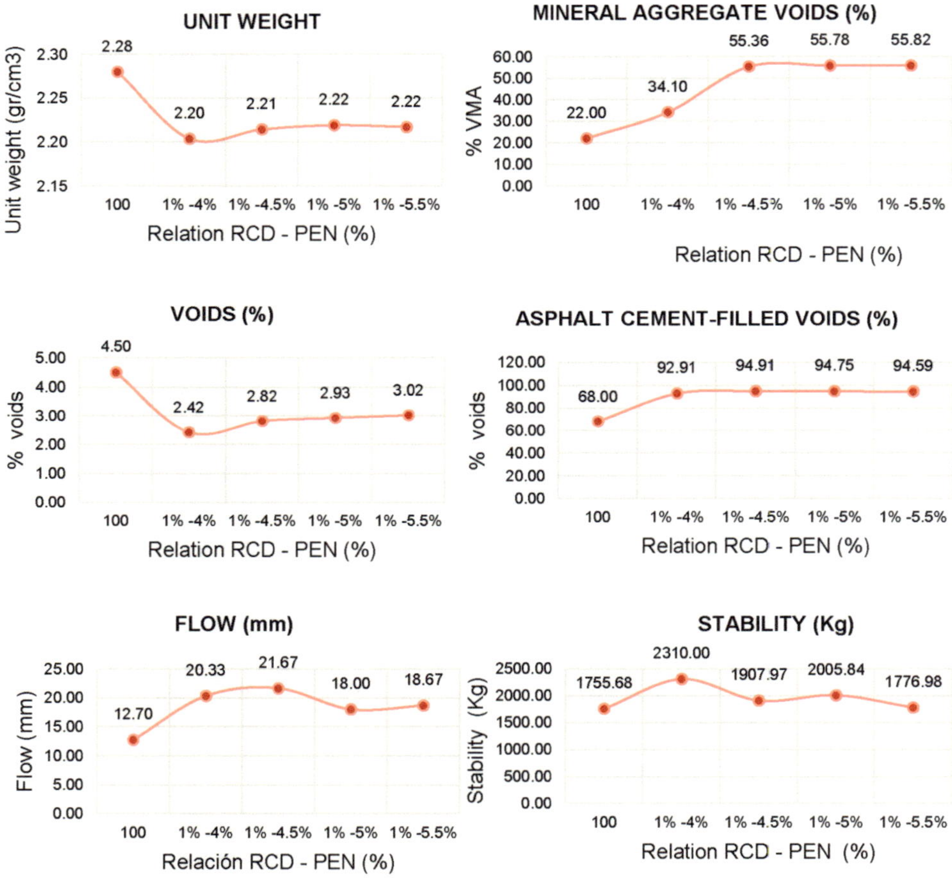

Table 9.
Analysis of the results of the mix design with 5%RCD - %PEN.

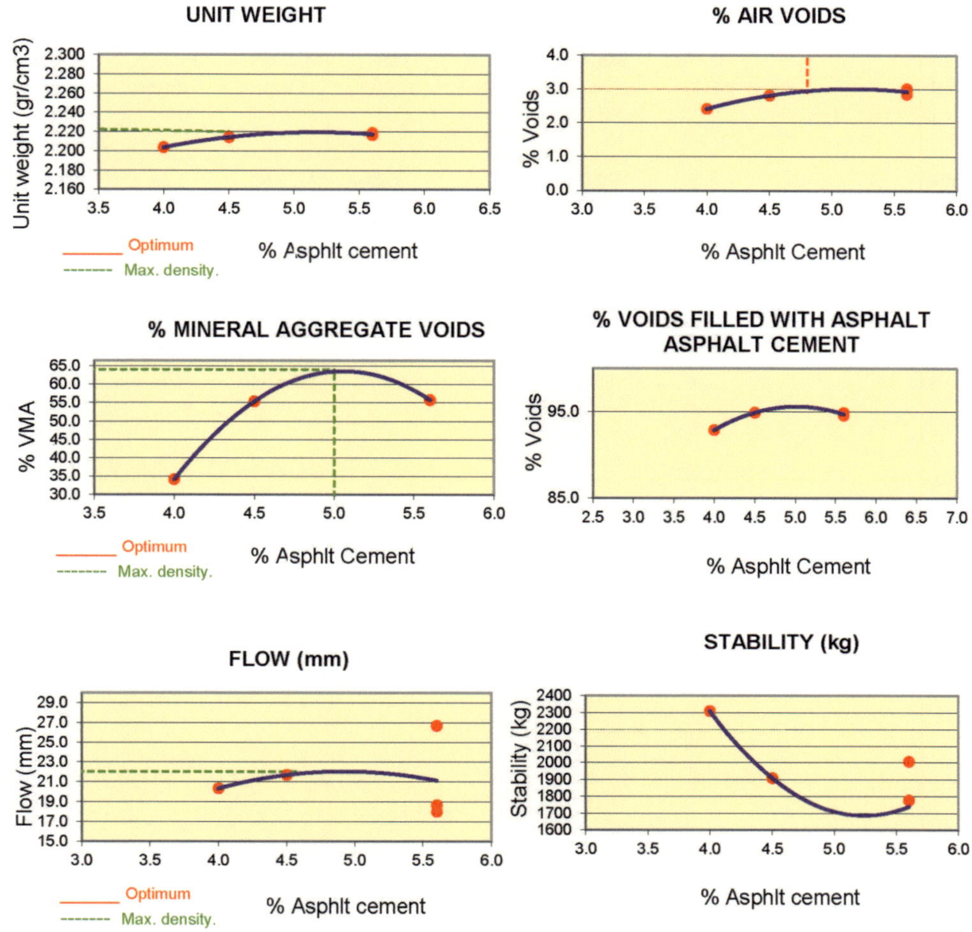

RESULTS	
Optimum Content C.A	4.97
Unit weight (gr/cm2)	2.23
Voids (%)	3.00
Mineral aggregate voids(%)	64.00
Voids filled with C.A (%)	96.00
Flow (mm)	5.59
Stability (Kg)	2300.00
Stiffness Index (kg/cm)	3250.00
Asphalt Dust Ratio	0.89

Figure 6.
Graphical representation of the asphalt design with 5%RCD.

Results								
Relation Pen-Filler	**Unit weight (gr/cm^2)**	**Voids (%)**	**Mineral aggregate voids (%)**	**C.A filled voids (%)**	**Flow (mm)**	**Stability (Kg)**	**Asphalt dust ratio**	**Flow stability ratio**
100	**2.28**	**4.50**	**22.00**	**68.00**	**12.70**	**1755.68**	**1.23**	**1809.22**
1–4%	2.19	2.50	36.22	93.14	13.00	2208.67	1.23	6708.03
1–4.5%	2.22	2.75	55.33	95.04	16.00	2184.99	1.23	5510.74
1–5%	2.22	2.85	55.84	94.90	22.67	2216.46	1.23	3855.23
1–5.5%	2.23	3.00	55.50	94.61	28.67	2060.48	1.23	2836.81
1–6%	2.23	3.01	55.59	94.60	24.00	2288.04	1.23	3863.21

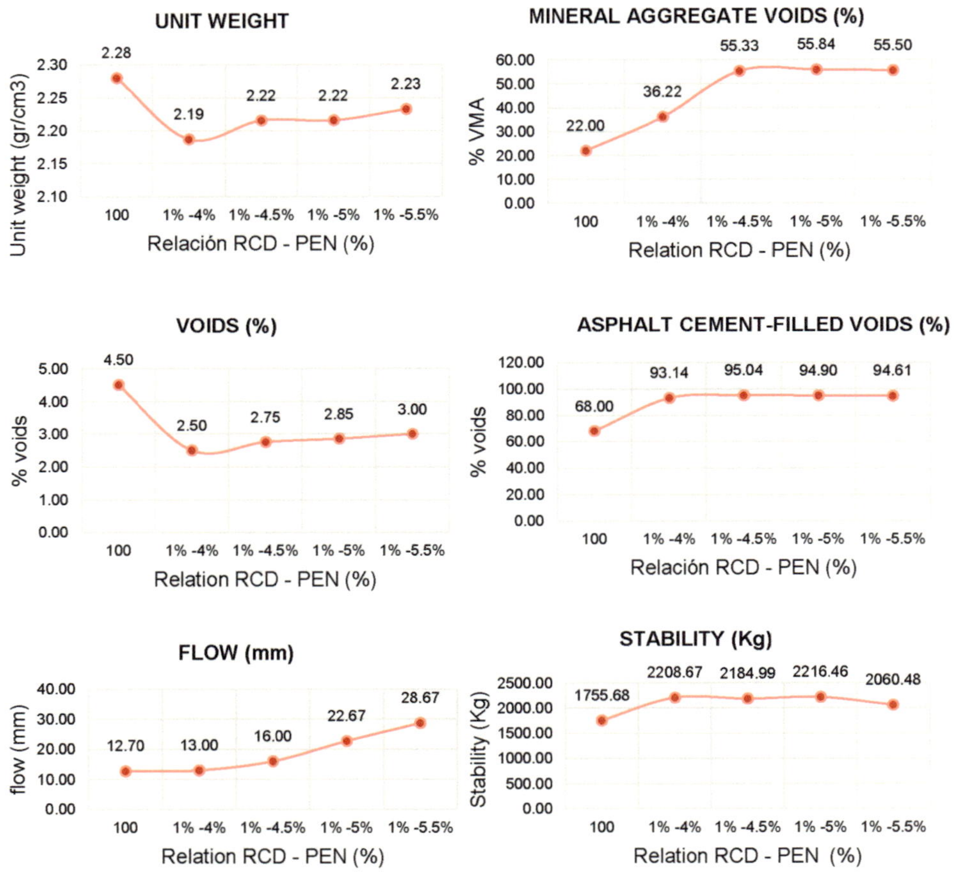

Table 10.
Analysis of mixture design results with 7%RCD - %PEN.

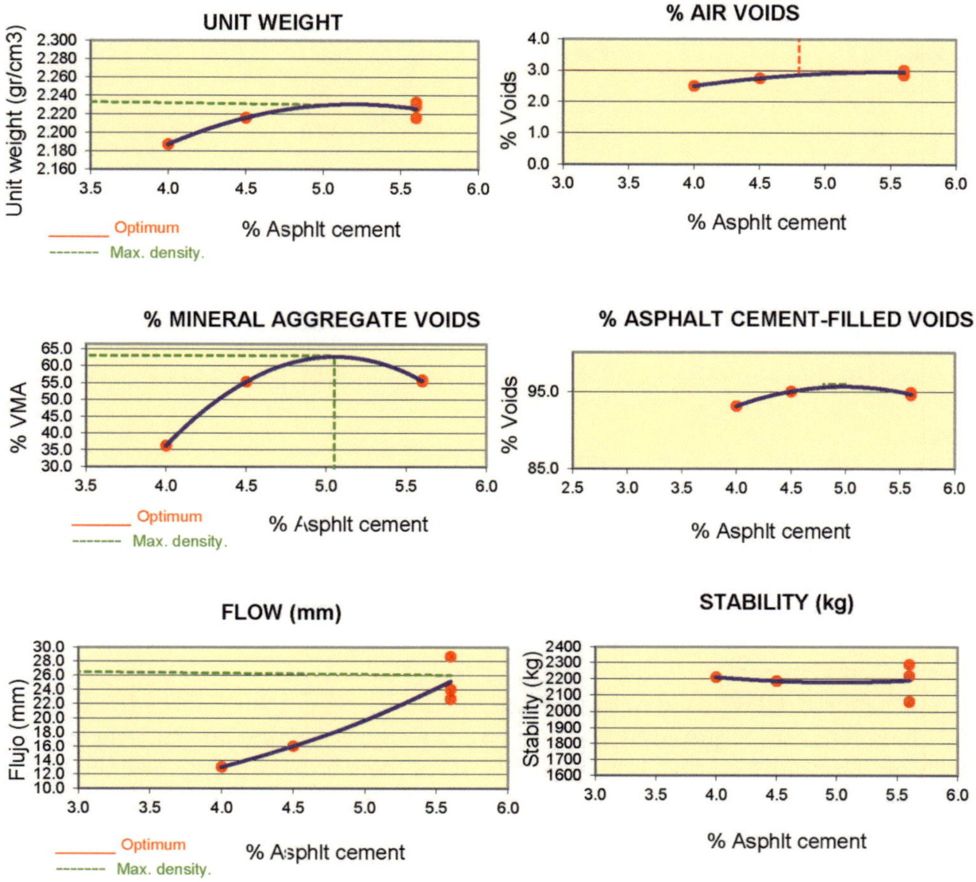

RESULTS	
Optimum Content C.A	4.77
Unit weight (gr/cm2)	2.24
Voids (%)	3.00
Mineral aggregate voids(%)	63.00
Voids filled with C.A (%)	96.00
Flow (mm)	6.86
Stability (Kg)	2300.00
Stiffness Index (kg/cm)	3250.00
Asphalt Dust Ratio	0.86

Figure 7.
Graphical representation of asphalt design with 7%RCD.

Comparison of asphalt design					
Filler	**1% Cement**	**1% CDW**	**3% CDW**	**5% CDW**	**7% CDW**
Optimum Content C.A	5.3	5.4	5.1	5	4.8
Unit weight (gr/cm^2)	2230	2300	2300	2230	2240
Voids (%)	3	2.8	2.6	3	3
Mineral aggregate voids (%)	65	65	61	64	63
Voids filled with C.A (%)	95	95	95	96	96
Flow (mm)	7.11	7.62	7.62	5.59	6.86
Stability (Kg)	1500	1800	2200	2300	2300
Asphalt Dust Ratio	0.95	0.98	0.92	0.89	0.86

Table 11.
Comparison results of the asphalt design with the optimal percentage.

4. Conclusions

The use of construction and demolition waste is an alternative ecological solution for the long-term sustainability of the construction industry for the reduction of environmental impact.

The concrete CDW used in the asphalt mix as filler has achieved the fineness and characteristics required by the standard through screening and the Angels machine. Through a comparison of results, it was concluded that the briquettes with the addition of concrete CDW filler and cement filler increase the stability being able to resist displacements and deformations, however, with the flow, it does not comply with the parameters having a minimum difference of deformation with cracking.

According to the additions of 1, 3, 5, and 7% of CDW of concrete, it was concluded that the addition of 1% of cement is similar to comply with the parameters according to regulations, while the addition of 5% of CDW of concrete only has greater resistance.

The addition percentages of 1%, 5% and 7% of RCD concrete meet the requirements established for the percentage of air voids. Regarding flowability, the 1% and 3% concrete RCD percentages showed the highest flowability among all the additions tested, while the 5% concrete RCD presented the highest stability. Regarding the powder-asphalt ratio and mineral aggregate voids, all the tested percentages comply with the parameters established in the Highway Manual. However, not all the filler percentages comply with the properties of the asphalt mixes, where stability is one of the most favorable results obtained, giving great resistance to displacement in the asphalt mixes.

Author details

Angel Antonio Ruiz Pico[1*], Ruth Noemi Salvador Reyes[2]
and Sol Estrella Espinal Bonilla[3]

1 Universidad Nacional Toribio Rodríguez de Mendoza de Amazonas, Chachapoyas, Peru

2 Universidad Católica Santo Toribio de Mogrovejo, Chiclayo, Peru

3 Fundación Angel Antonio Ruiz Pico, Chiclayo, Peru

*Address all correspondence to: aaruizpico@gmail.com

References

[1] Pastrana-Ayala J, Silva-Urrego Y, Adrada-Molano J, Delvasto-Arjona S. Propiedades físico-mecánicas de concretos autocompactantes producidos con polvo de residuo de concreto. Informador tecnico SENA: Servicio Nacional de Aprendizaje. 2019. Available from: https://docs.google.com/document/d/11TCPqbpHX_Q5io4otNKxxd7HeaA3q_Go/edit?usp=drive_link&ouid=108936756070772477317&rtpof=true&sd=true

[2] Aparicio P, Martína D, Baya-Arenas R, Flores-Alés V. Behaviour of concrete and cement in carbon dioxide sequestration by mineral carbonation processes. Ceramica y Vidrio. 2022

[3] Maleki Toulabi H, Hossein SA, Najaf LH. Investigating the effect of using recycled waste concrete powder (RWCP) on the performance of thin layer surface treatment: Microsurfacing. Construction and Building Materials. 2023

[4] Flores P, Flores M, Flores Victor R, Torres R, Mairon K. Mezclas asfálticas con materiales reciclados de construcción y demolición para la reparación de pavimentos. Revista de Ingeniería Civil ECORFAN-Perú. 2020

[5] Silva-Urrego Y, Delvasto-Arjona S. Uso de residuos de construcción y demolición como material cementicio suplementario y agregado grueso reciclado en concretos autocompactantes. Informador Técnico. 2021

[6] Gedik A. Una revisión sobre la evaluación del aprovechamiento potencial de residuos de construcción y demolición en pavimentos de mezcla asfáltica en caliente. Resources, Conservation and Recycling. 2020

[7] Acevedo, Agudelo H, Figueroa Álvarez J. Prácticas de circularidad en la gestión de residuos de construcción y demolición en la industria de la construcción: una revisión de la literatura sobre las estrategias y aspectos clave sobre su implementación. Informes de la Construcción. 2023

[8] Martínez G, Caicedo B, González D, Fuentes L, Torres V. Trece años de desarrollo continuo en mezclas asfálticas modificadas con caucho granulado en Bogotá: logrando la sostenibilidad de los pavimentos. Revista Ingeniería de Construcción. 2018

[9] Bastidas-Martínez JG, Rondón-Quintana HA, Contreras-Zartha L, Forero-Castaño S, Rojas-Rozo L. Evaluación de una mezcla de concreto asfáltico con incorporación de agregados reciclados de concreto. Revista UIS Ingenierías. 2021

[10] Bravo J, Valderrama C, Ossio F. Cuantificación Económica de los Residuos de Construcción de una Edificación en Altura: Un Caso de Estudio. Información Tecnológica. 2019

[11] Arroyo P, Herrera R, Salazar L, Martínez J, Calahorra M. Un nuevo enfoque para integrar factores ambientales, sociales y económicos para evaluar mezclas asfálticas con y sin neumáticos de desecho. Revista Ingeniería de Construcción. 2018

[12] Radevic A, Đurekovic A, Zakic D, Mladenovic'a G. Effects of recycled concrete aggregate on the stiffness and rolling resistance of asphalt concrete. Construction and Building Materials. 2017

[13] Acosta Alvarez D, Martínez RM, Guerra GG. Influencia de la utilización

del CDW como árido en mezclas asfálticas en caliente. Revista Arquitectura e Ingeniería. 2016

[14] Loaiza A, García E. Evaluación de asfalto mezclado con polvo de corteza y fibras residuales de coco para aplicaciones estructurales. Revista de la Construccion. 2019

[15] Espino Gonzales CU, Lara Gomez C, Trujillo Calderón I, Nuñez Guzmán EA, Tafolia Martinez E. Elaboración de mezclas asfálticas en caliente con adición de residuos de construcción y demolición (concreto hidráulico) en un 10% y 20% como sustitución de agregado pétreo natural. XV Congreso Latinoamericano de Patología de la Construcción y VII Congreso de Control de Calidad en la Construcción. 2019

[16] Espino González CU, Molina WM, Guzmán EMA. Elaboración de mezclas asfálticas en caliente con adición de residuos de construcción y demolición como sustitución de agregado pétreo natural. Ciencia Nicolaita. 2023

[17] Pacheco Bustos CA, Fuentes Pumarejo LG, Cotte ÉHS, Quintana HAR. Residuos de construcción y demolición (CDW), una perspectiva de aprovechamiento para la ciudad de barranquilla desde su modelo de gestión. Ingeniería y Desarrollo. 2017

[18] Zoua G, Zhanga U, Liub X, Lina Y, Yua H. Design and performance of emulsified asphalt mixtures containing construction and demolition waste. Construction and Building Materials. 2020

Chapter 2

Building Information Modeling (BIM) Implementation and Practices in Construction Industry: A Review

Bamidele Charles Olaiya, Olaolu George Fadugba and Mustapha Muhammad Lawan

Abstract

In the construction sector, Building Information Modeling (BIM) has become a disruptive technology that improves project outcomes and changes conventional methods. This analysis looks at how BIM is used in the construction industry, emphasizing the advantages, factors that encourage adoption, and obstacles to it. Better stakeholder cooperation and communication are made possible by BIM, which boosts project efficiency and saves a substantial amount of money. Its value extends beyond the building phase into facilities management by supporting sustainable construction practices through thorough energy analysis and lifecycle management. Despite these benefits, there are still obstacles to BIM adoption, including high upfront costs, shortage of workers with the necessary skills, interoperability problems, and organizational opposition. The degree of BIM adoption throughout the world varies depending on elements such as industry norms, government regulations, and technology readiness. Industry stakeholders must work together to define global standards, provide training and education, create supporting regulations, and share successful case studies in order to overcome these obstacles. BIM is positioned to play an increasingly important role in the construction industry's ongoing evolution, spurring innovation and enhancing the built environment.

Keywords: building information model, construction industry, technology, global trends, communication

1. Introduction

Building and infrastructure facility development and implementation are intricate processes involving several parties with varying levels of skill [1]. These parties must constantly reconcile and communicate intensely in order for the building project to be successful. As of right now, this usually entails the delivery of the construction project's technical drawings in a graphical format, such as sections, views, and detail drawings that are both horizontal and vertical [2]. Digitalization has drastically increased productivity, product quality, and product diversity during the previous 10 years in a number of industrial sectors. Digital technologies are being used more often in the Architecture, Engineering, and Construction (AEC) sector to design, develop, and manage infrastructure assets and buildings [1, 3, 4]. However, compared to other business fields, the continuous usage of digital information throughout the whole production chain lags substantially. Because information is still mostly transferred in the form of drawings either as actual printed plots on paper or in a restricted digital format, valuable information is lost much too frequently. These interruptions in the flow of information happen at every stage of a constructed facility's lifespan, including the crucial handovers that take place during the phases of design, building, and operation [5]. Building information modeling can be useful in this situation. A much deeper application of computer technology in the engineering, design, building, and operation of constructed facilities is accomplished through the use of the BIM approach [5].

The term "Building Information Modeling" (BIM) is most commonly used to describe a collection of parametric tools and processes for the creation and upkeep of an integrated collaborative database of multi-dimensional information about the planning, construction, and operation of a building. The goal of BIM is to improve stakeholder collaboration, which in turn shortens the time required for project documentation and yields more predictable project outcomes [2, 5].

Planning, designing, constructing, and managing infrastructure and buildings may be revolutionized with the use of Building Information Modeling (BIM). Building Information Modeling (BIM) is the process of digitally capturing a building's structural and operational features. It establishes a common knowledge base for building-related information and provides a solid foundation for decision-making throughout the building's lifecycle, from conception to destruction. Another way to describe building information modeling (BIM) is as an intelligent model-based method that offers knowledge for planning and directing infrastructure and construction projects more quickly, cheaply, and effectively [6, 7]. Building Information Modeling (BIM) is a process that includes creating and maintaining digital representations of a place's functional and physical attributes. It is not only a technology. In order to create and convey project choices, it entails building and utilizing sophisticated 3D models. Due to the abundance of data in these models, comprehensive simulations and analyses of building performance, scheduling, and cost control are possible.

The origins of BIM may be found in the early years of computer-aided design (CAD) in the 1970s and 1980s. The idea has developed over several decades. However, the early 2000s saw the popularization of the phrase "Building Information Modeling" itself. Progress in processing power, software capabilities, and industry standards has improved BIM's functionality and uptake over time [7]. BIM is a technology-enhanced approach that brings together a variety of stakeholders from many professions to work in a multidisciplinary setting, exchanging information and cooperating on a single project that is constructed more quickly, more cheaply, and with fewer mistakes [8, 9]. Beyond basic spatial connections, BIM goes beyond standard CAD drawings by offering intelligence to individual building components (such as windows, walls, or chillers) as well as system- and building-wide information and awareness (such as system flows or building loads).

BIM models provide a wealth of information that simulates the whole construction cycle, including but not limited to 3D geometry, procurement, various building operations, costs, and time schedules. The models that have been developed enable the extraction of data on quantity take-offs, materials needed, and task scope. The three primary goals of every construction project, i.e., increased quality, shortened time, and lower cost, can be accomplished with the quantity of information shared and comprehended [8, 10]. The following procedures are included in building information modeling [9]:

- Creating and utilizing a system of integrated graphical data management and information flow in relation to the construction process description;

- Developing an integrated and holistic building creation strategy encompassing design, construction, and lifecycle management based on modeling and computer simulation; and

- Converting lone contractors into teams that function as decentralized units that tackle complex problems and integrate disparate tasks into coherent processes.

It is anticipated that these BIM procedures would result in increased productivity and decreased expenses for different operations over the course of the building's lifetime [11]. Architects, engineers, contractors, owners, facilities managers, and other project lifecycle participants are all involved in the BIM process and communicate with BIM designers, who are tasked with providing more precise energy modeling data. Then, BIM uses this data to create and utilize digital models for building project design, construction, and/or operation. These models create a virtual building database that serves as a single, integrated source for all information related to the design of a building by combining intelligent 2D and 3D objects used to define a building design with external factors like location and local conditions [8, 11].

Architects, mechanical, electrical, and plumbing (MEP) engineers, and contractors are able to depict the geometric and functional relationships between building elements because of the parametrically defined graphical and non-graphical information that is attributed to the objects. All of the building project's design papers and timetables are fed by this data into an integrated database [11, 12].

1.1 Core components of BIM

- *3D modeling*: At the heart of BIM is the creation of detailed three-dimensional digital models that represent the geometry of a building. These models are built using parametric components, which means that any change in one part of the model automatically updates related parts, maintaining consistency and accuracy.

- *Information management*: Information is managed over a building's whole lifespan as part of BIM. Design data, construction timetables, cost estimates, and maintenance data are all included in this. BIM enables improved communication and cooperation among project stakeholders by centralizing this data.

- *Collaboration and integration*: BIM encourages owners, contractors, engineers, and architects to work together. BIM lowers mistakes and rework by offering a common platform where all stakeholders can view and update project information, resulting in more effective project delivery.

- *Lifecycle management*: BIM's emphasis on a building's complete lifespan, from original design to construction to operation and maintenance, is one of its fundamental advantages. With this lifetime strategy, you can be confident that the building will function as planned and can be effectively managed during its usage.

1.2 Benefits of BIM

The main advantages of BIM include the following:

a. *Flow of information*: One of the primary benefits of BIM is that it may greatly enhance the flow of information at every step of a structure's design, construction, and lifecycle. This is because a digital model is a unified description of a building. At various points throughout the building lifecycle, the architects, MEP engineers, contractors, facilities managers, and owners may use the digital model to add, remove, or change information to support their roles. Maintaining a clear project vision helps to increase productivity, decrease mistakes, and encourage well-informed decision-making [13].

b. *Improved design visualization*: BIM gives stakeholders access to a thorough 3D model of the building, enabling them to see the project before it is constructed. This facilitates the early detection of any design problems and the formulation of wise decisions. For architects and engineers, anticipating the reactions and interactions of building residents, guests, and neighbors is an essential aspect of the design process. Nonetheless, MEP engineers may also gain a great deal from the virtual building model produced during the BIM process, which aids in optimizing the HVAC system architecture within the architectural restrictions of the building [13].

c. *Better cost estimation*: Due to the comprehensiveness and accuracy of the data it offers, BIM can streamline and aid in the creation of more accurate cost estimates. The relative simplicity with which assembly and material amounts may be retrieved from the model can speed up and improve the accuracy of estimates, giving a clearer indication of the effect of design modifications and enabling proactive management of budgetary problems [10, 14].

d. *Lower construction costs*: One of the numerous BIM tools, clash detection, may be used to indicate where elements of the design share space long before the building even starts. This may lessen or completely remove the requirement for field modifications while work is underway. Additionally, BIM models may be utilized to confidently prefabricate building components like pipe or duct lines. This can reduce the price of installation and assembly. Additionally, BIM supports financial planning and budgeting by offering precise quantity take-offs and cost estimations. Time and waste are also decreased, and building timelines may be optimized by having the capacity to identify conflicts.

e. *Building history*: For owners and service contractors, the digital model may operate as a valuable information repository as a building moves through the design, construction, and occupancy phases. The building information model, for instance, may be used to locate and identify the manufacturer, model number, performance standards, and other relevant information about a malfunctioning

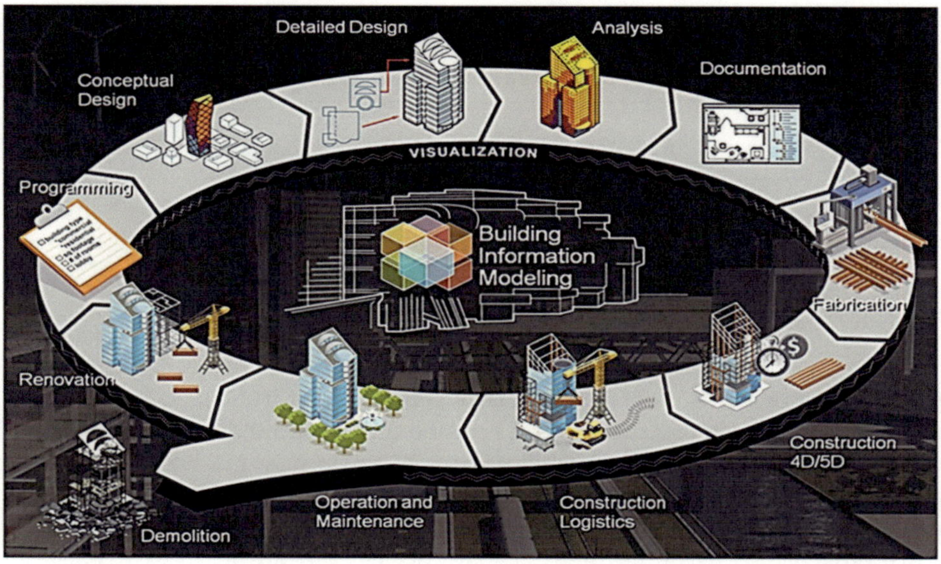

Figure 1.
Schematic representation of BIM interaction [15].

building component so that it can be replaced or repaired as quickly as possible. The building information model may be used to uncover hidden elements like electrical equipment, ductwork, and plumbing if a section of the structure is being rebuilt. This can help with decision-making on the remodel plan [1].

A schematic depiction of how BIM is used in the building process is shown in **Figure 1**.

2. Characteristics of BIM

2.1 BIM competence stages

The fundamental capacity to carry out a job or provide a BIM service or product is known as BIM capability. The minimal BIM criteria, or BIM Stages, are the significant benchmarks that teams or organizations must meet while putting BIM technologies and concepts into practice [8]. The BIM Capability model, which ranges from level 0 to level 3, is a type of model used to assess stakeholders' software proficiency. It is intended for use by individuals or organizations engaged in BIM. Three phases of BIM capability exist, and they are [8]:

- Object-based modeling in BIM—Stage1;
- Collaboration based on models—Stage 2;
- Integration based on networks—Stage 3.

A BIM stage's minimal criteria describe it. For instance, an organization must have implemented an object-based modeling software product, such as ArchiCAD, Revit,

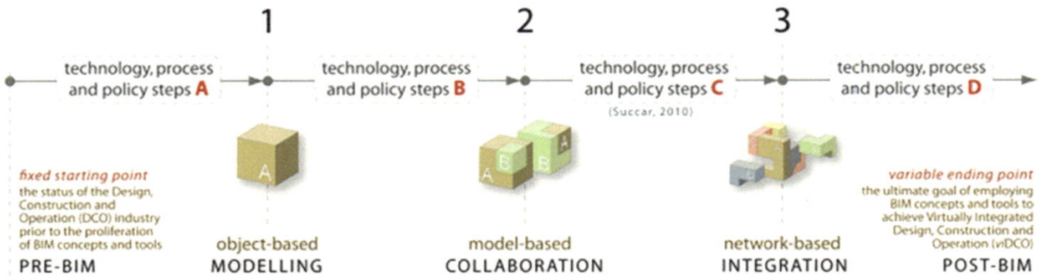

Figure 2.
BIM capability stages [8].

Tekla, or Constructor, in order to be evaluated at BIM Capability Stage 1. In a similar vein, an organization must participate in a collaborative interdisciplinary "model-based" project in order to achieve BIM Capability Stage 2. An enterprise must be sharing object-based models with at least two other disciplines via a network-based solution (such as model servers) in order to be evaluated at BIM Capability Stage 3. Competency Steps are further broken into each of the three Capability Stages. Steps are progressive changes, but stages are transformative or drastic ones [16]. This is what sets stages apart from steps. The set of actions necessary to go through a BIM Stage, from pre-BIM to post-BIM, is determined by the unique requirements, obstacles, and deliverables associated with each stage. The three BIM capability phases are depicted in **Figure 2**.

2.2 BIM maturity levels

The quality, consistency, and level of excellence within a BIM capability are referred to as "BIM Maturity." That is to say, "maturity" refers to the degree of that skill in carrying out a job or delivering a BIM service or product, as opposed to "capability," which indicates minimal competence. Benchmarks for BIM Maturity are performance improvement levels or milestones that organizations and teams strive toward. An organization's hybrid mix of humans and technology needs to be of a high caliber in order for it to be classified as very mature in BIM. This high quality translates into better control over differences in time schedules and cost estimates between goal and actual results [8, 17]. The shift from low to higher levels of maturity often indicates:

- Greater efficacy in achieving predetermined objectives and establishing new, more ambitious ones;
- Better control by eliminating differences between performance targets and actual outcomes; and
- Better predictability and forecasting by lowering variability in competency, performance, and costs [11].

It is not possible for the construction sector to make the significant shift to completely digitalize model-based working methods all at once. A better strategy would be to gradually implementing the new technology together with the related process modifications. Consequently, the BIM Maturity Model was created by the UK BIM Task Group and outlines four distinct stages of BIM deployment, which are [18] as follows:

1. *Level 0 (Pre BIM)*: This is the initial phase, the state prior to the use of BIM, and it denotes an unmanaged CAD environment with no project team participation. Level 0 refers to customary workflows that utilize 2D computer-aided design (CAD) and include exchanging paper-based designs. Information sharing most likely occurs via 2D documents. While 2D is the foundation for all documents, 3D visualizations are also useful. In general, the visualization model and documentation are not connected to quantities, cost estimates, and requirements. There is no digital collaboration at this point. Paper drawings, computer prints, or a combination of the two are the results.

2. *Level 1 (Object-based modeling)*: Models are created using an object-based 3D parametric software tool. All phases of creation include users creating models, which serve as the foundation for both 2D and 3D documentation and visualization. It consists of the facility's partial 3D modeling, particularly for complicated geometries, with the majority of the design still being realized through 2D drawings. Individual files are sent and received in this case to facilitate data interchange; a central project platform is not used.

   Stages 0 and 1 differ from each other in terms of contractual relationships and small process modifications. A combination of 2D and 3D CAD is usually used in stage 1 for documentation and product information, and for concept work.

3. *Level 2 (Collaborative BIM)*: Using a model-based collaboration method, several disciplines actively collaborate with one another at this level. The cloud-based application is typically used for collaboration. It becomes important to alter contracts, and different analytical tools might be coupled to models. The models are used to evaluate processes. Rather than using a single, common model, each field is developing its own model. The key component of this level is the information sharing throughout disciplines, which is how cooperation emerges. Within a Common Data Environment (CDE), design information is shared using standardized formats such as Industry Foundation Classes (IFC) or Construction Operations Building Information Exchange (COBie). This makes it possible for businesses to integrate data with their own model and apply the knowledge going forward.

4. *Level 3 (Integrated BIM)*: During the course of the project lifecycle, jointly developed, shared, and maintained semantically rich network-based integrated models are produced. At this point, models become multidisciplinary models that permit complicated analyses to be done during the early phases of design and construction. Deliverables from the model include full lifecycle costing, green policies, business information, and lean building techniques. Significant adjustments to contractual arrangements, risk distribution schemes, and procedural procedures are now required. In order to offer two-way access to project stakeholders and ultimately enable Integrated Project Delivery (IPD), a common transdisciplinary model is required. The whole lifespan makes use of tightly integrated digital models, and ISO standards are used for data interchange and process descriptions. Project data is managed using cloud services to ensure that it is regularly and continually maintained throughout the building's lifecycle. An illustrated depiction of the BIM maturity stages is presented in **Figure 3**.

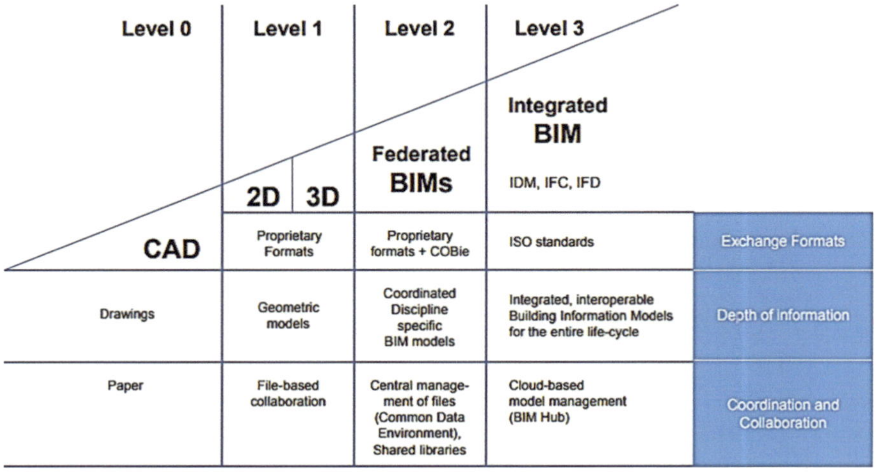

Figure 3.
The BIM maturity ramp [19].

2.3 BIM competency sets

For the purposes of BIM deployment and evaluation, a hierarchical collection of individual capabilities is known as a BIM Competency Set [19, 20]. The word "competence" refers to a general set of skills appropriate for implementing and evaluating BIM capability and/or maturity rather than necessarily reflecting human talents. They are known as BIM Implementation Steps if a BIM Competency Set is applied to active implementation. On the other hand, they are known as BIM Assessment Areas when they are utilized to evaluate current implementations. BIM competencies may be categorized into three categories: technology, process, and policy. These sets are a direct reflection of BIM requirements and deliverables.

Networks, hardware, and software are examples of technology. For instance, the transition from drafting-based to object-based workflow is made possible by the availability of a BIM tool, which is a prerequisite for BIM Stage 1. Process sets for infrastructure, human resources, leadership, and goods and services. Model-based collaboration, for instance, requires database-sharing abilities and collaboration procedures (BIM Stage 2). Contracts, rules, and research/education all contain policy. Network-based integration, for instance, requires alliance-based or risk-sharing contractual arrangements (BIM Stage 3).

2.4 BIM level of development (L.O.D)

In Building Information Modeling (BIM), the Level of Development (LOD) indicates the degree of precision and detail that a model element has at different points during the project lifecycle. Clarity, fewer misunderstandings, and improved decision-making are all aided by the standardized method that LOD offers for describing and conveying the dependability and content of BIM parts [21]. Usually, there are 500 or more LOD frameworks, each with unique properties and uses. Depending on the magnitude of the project, the client's requests from the consultants and contractors determine what LOD should be employed [22].

- *LOD 100 (Conceptual)*: Elements are visually depicted but lack detail at this level. In essence, they are placeholders that indicate general dimensions, form, position, and orientation.

- *LOD 200 (Approximate geometry)*: Non-geometric data may be included in elements that are defined by approximate geometry. It is appropriate for schematic design phases when preliminary design intent is expressed, and it includes basic material requirements and other non-geometric data.

- *LOD 300 (Precise geometry)*: Element dimensions and geometry are precisely represented at this level. These components can be used to construct detailed designs and to coordinate with other professions. It contains comprehensive non-geometric details on the finishes, materials, and performance requirements.

- *LOD 350 (Detailed design)*: All of the elements, including interconnections with other building systems, are designed with intricate geometry and linkages. It is utilized for thorough coordination, conflict detection, and construction documentation and contains precise information regarding fabrication and installation.

- *LOD 400: (Fabrication and assembly)*: At this level, particular fabrication, assembly, and installation details are represented for each element. It contains details on how to assemble, install, and determine exact measurements.

- *LOD 500 (As-built)*: The as-built circumstances are represented by the elements that are modeled. Verified information representing the finished construction is included at this level. It has extensive data for facility management, operations, and maintenance.

In BIM, the Level of Development (LOD) is a crucial framework that establishes the precision and level of detail of model pieces during the course of the project. Project teams may improve their clarity, communication, and decision-making, which will result in more effective and successful project outputs, by comprehending and applying LOD. From conceptual design to as-built documentation, each step of the lifecycle (LOD) has a distinct function that supports the various demands of various project phases [22].

2.5 BIM dimension

The many information and analytical facets that may be included in a BIM model are represented by the dimensions of building information modeling (BIM) [23, 24]. These dimensions encompass extra data that may be utilized to improve the design, building, and management processes, going beyond the fundamental 3D modeling of physical structures. These are the BIM dimensions that are widely accepted [24, 25]:

- *1D BIM*: The documenting of all needs related to the building project lifecycle is the first basis of BIM. The BIM platform incorporates the scattered information from stakeholders and teams as a foundation for managing configuration and document updates. The process of creating and disseminating crucial information throughout the infrastructure building project's commissioning phase is made clearer by documentation. With consistent access to engineering specifications, architectural drawings, equipment features, and product requirements, the structured set of project documents helps project managers exchange information with the project team and keep them on track. It also streamlines the information management process [26]. Processing changes and maintaining track of various document versions need proper document management. The document management system defines which internal and external stakeholders have real-time access to a complete archive of documents. Owners may obtain a comprehensive understanding of the distribution of information among stakeholders and the dimension of information by centralizing all crucial papers required for the design, execution, and administration of a facility. This BIM feature simplifies a number of steps in the infrastructure construction process.

- *2D BIM*: Project design and drawings can only be represented using a basic X- and Y-axis model when modeling a project in two dimensions. Planning is mainly done in two dimensions, connecting the project requirements to the goals and constraints. Because 2D BIM is the most ancient type of construction model, it makes basic planning tasks faster and easier to understand. More detail is necessary, nevertheless, in big and complicated infrastructure projects to guarantee the creation and implementation of perfect, cost-effective plans and designs that enable the timely completion of building projects. Detailed planning is more difficult when more variables and restrictions are added, and visualizing the parameters becomes necessary [27].

- *3D BIM*: Three-dimensional items make up the information model in a three-dimensional BIM model. In a virtual world, these things symbolize the building or building areas. These three-dimensional items have at least the following dimensions: length, breadth, and height. Additional details, including the material and finish, can be used. The clarity and rigor of the design and planning process are increased when done in a three-dimensional setting. It comprises the integration and representation of both graphical and non-graphical data, including estimated values, spatial connections, and isometrics. A 3D information model's main advantages are enhanced coordination, made possible by visualization, and generic information collection. Designers might enhance the quality of the results by simulating potential physical collisions during the assembly of various components. This type of quality assurance for design documents helps to enhance project compliance with quality standards by eliminating mistakes. Any upgrades, including new construction, modifications, and destruction, may be handled more methodically [28].

- *4D BIM*: Time is a dimension that 4D BIM adds to the 3D model. In order to replicate the building process across time, this entails incorporating construction scheduling and sequencing into the BIM model. When a facility's schedule is integrated into its 3D model, timing and activity sequence issues may be identified. Project schedules must be closely examined for any apparent problems with conflicts and interferences in the hard or soft logic of the activity relationships. Using the BIM analytical tools, the planned activities' evolution is simulated so that the activity network may be enhanced and optimized.

- *5D BIM*: Cost data is incorporated into the BIM model using this BIM dimension. This is tying cost information to the different parts of the 3D model so that financial analysis, budgeting, and cost estimation may be made easier. The seamless integration of budget, schedule, and BIM 3D models allows estimators to assess capital and operational expenses during the building phase and gives the owner precise information about how much the contractor should be invoicing at any given time. Automated quantity surveying may be used to achieve a realistic budget by visually analyzing the sensitivity of the expenses associated with carrying out each operation over time [29]. The primary advantages of the fifth dimension of BIM may include increased project change accuracy and predictability as well as a more trustworthy cost estimate of various construction scenarios.

- *6D BIM*: Energy efficiency and sustainability are the main topics of 6D BIM. This dimension entails adding information to the BIM model on environmental effects, sustainability initiatives, and energy analysis. It makes it possible to find out at a very early (concept) stage information on the building's anticipated energy usage.

- *7D BIM*: The operations and management of facilities are the focus of 7D BIM. In this level, data on the building's upkeep, operation, and lifecycle management are included. The data comprises asset characteristics, information on operation and maintenance during the project commissioning phase, facility requirements, installation and warranty details, maintenance schedules, manuals, and equipment configurations required for optimum performance. Owners can utilize this data to optimize infrastructure maintenance and operation in order to meet sustainability goals. With the aid of the 7D BIM, managers can see the lifecycle costs of their facilities and make well-informed decisions that take into account all potential long-term effects on the development or modifications of the facility. Through the utilization of lifetime data, designers are able to take Total Cost of Ownership (TCO) into account while developing infrastructure [30]. Some frameworks suggest other dimensions in addition to the widely accepted ones to solve other BIM-related issues:

- *8D BIM (Safety)*: Safety management and planning are incorporated into the BIM process using 8D BIM. It entails adding risk analyses and safety precautions to the model. This dimension facilitates smooth communication and interaction between managers and stakeholders in order to implement safety strategies from the very beginning of the facility's lifespan. This BIM system feature has not been completely utilized in reality, though, and more efficient hardware and software are still required to complete this integration. 8D BIM could only identify and remove the safety hazard by visually analyzing the building and all of its parts [31].

- *9D BIM*: 9D BIM emphasizes the concepts of lean building. It entails streamlining the building process to reduce waste and increase productivity [32]. This dimension focuses on resource management strategies to enhance the distribution and utilization of labor, materials, tools, and equipment across the facility's lifetime. All resources used in the building and maintenance of infrastructure are analyzed using 9D BIM. For instance, helpful information may be obtained on how best to employ trucks for material transportation, cut down on the amount of onsite cars and circulation routes, get rid of repetitive, non-value-adding work, and shorten cycle times.

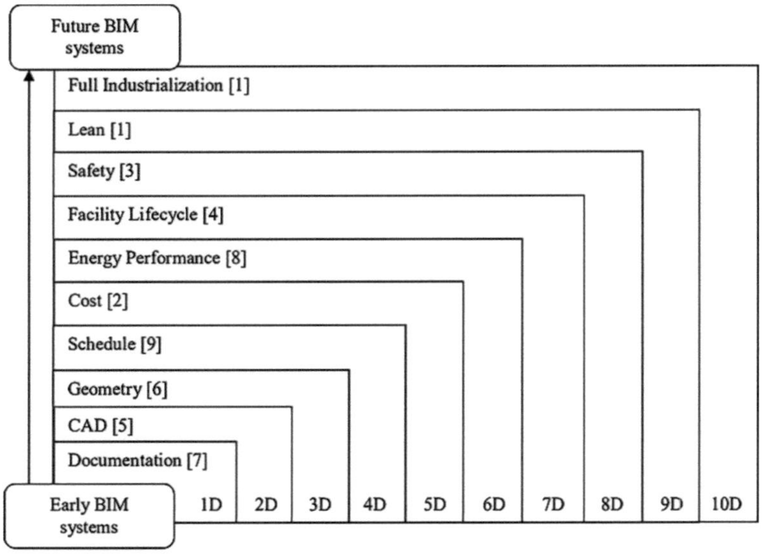

Figure 4.
Proportions of the BIM system [33].

- *10 D BIM*: Another potential addition to the BIM system is the 10D BIM, which integrates disaster management plans and attempts to capitalize on industrialized construction [32]. This dimension finds and removes impediments to productivity in the planning, building, and operation of a facility. This dimension promotes the use of manufacturing equipment and drones to increase production. In this field, artificial intelligence is crucial for automating engineering planning and control processes. This dimension was very recently established, therefore, further research and testing are needed to determine how to use it. By using instantaneous information management, infrastructure building may be made more rigorous by including a higher degree of automation and systematic control, which also reduces the negative environmental effects.

The many BIM dimensions are displayed in **Figure 4**, along with the tasks to which each dimension was assigned in order to achieve optimal performance.

In summary, BIM dimensions offer a thorough framework for incorporating different kinds of data and analysis into a BIM model. These dimensions provide improved construction project planning, coordination, and administration by going beyond 3D geometric modeling. Throughout the building lifecycle, better project results, more efficiency, and increased sustainability can result from an understanding of and use of the various BIM elements.

3. BIM application at various construction phases

3.1 Design development phase

In order to produce a thorough and well-coordinated collection of drawings and specifications, the preliminary design concepts are improved and detailed during the

Design Development (DD) phase, which is an essential step in the building process. During this phase, Building Information Modeling (BIM) is essential since it improves accuracy, cooperation, and efficiency. There are several benefits that BIM offers to the engineering and design process. One of the biggest benefits of adopting BIM over traditional 2D procedures is that most technical drawings, including horizontal and vertical sections, are generated straight from the model and are thus automatically compatible with one another. Early conflict identification and resolution between design disciplines is made feasible by clash detection between the various partial models [23, 34]. Because a lot of the input data on the geometry and material properties of the building can be obtained straight from the model, BIM also makes it easier to integrate computations and simulations in an efficient manner. The design process can then make use of a variety of simulations, such as lightning analysis, evacuation simulations, building performance simulations, and structural analyses. Furthermore, the model may be used to verify that it complies with rules and laws; it is mostly semi-automated at the moment, but it may eventually be more automated. Lastly, a highly accurate quantity take-off may be computed using the model data, which serves as the foundation for trustworthy cost estimates and enhances accuracy during the tendering and bidding process [31].

The majority of the design and engineering work in traditional planning procedures happens in the latter stages of detailed design, and occasionally even during the building phase itself. Because of this, a thorough evaluation of the building design happens only at a very late stage in the process, along with the intricate coordination of design disciplines and the integration of analytic and simulation technologies. However, at this time, there are fewer and more expensive implementation options for design modifications [34].

In contrast, a large portion of the planning work in a BIM-based planning process can be advanced to the early stages of design by creating an extensive digital building model. It is possible to evaluate the impact of design decisions more thoroughly and to identify and resolve potential conflicts early on by using computational analyses and meticulous planning of coordination requirements in the early design phases. This reduces the effort required in later phases and improves the overall quality of the design [26]. The main advantages of applying BIM throughout a building project's design and development phase are as follows [29]:

- Concepts become clearer and project conceptualization is made easier.
- The owner is given a more straightforward and accurate visualization of the design.
- Design decision-making is supported.
- Feasibility studies, simulations, coordination, and design quality are improved.
- Design and installation conditions are improved.
- Design time and cost are saved.
- Mistake identification is improved, reducing conflict error.
- Collaboration among construction players is improved.

- Risk reduction and an improved planning process are achieved.
- Early on, more consideration is given to the selection of construction components.
- BIM identifies potential conflicts before construction begins.

The use of BIM greatly improves accuracy, coordination, and stakeholder participation throughout the Design Development phase of the project. Detailed modeling, conflict detection, performance simulations, and thorough documentation are made easier by BIM procedures and technologies, which improve project outcomes and help designers make better decisions. Construction projects can gain greater quality, more productivity, and better alignment with client and regulatory requirements by utilizing BIM throughout this phase [29].

3.2 Construction phase

Throughout the building process, Building Information Modeling (BIM) is an essential tool, and one of the most important phases when its advantages are felt is during the construction phase. BIM facilitates better project execution, increases productivity, and guarantees improved quality control throughout this stage [35].

The utilization of Building Information Modeling (BIM) has noteworthy benefits for the planning and implementation of a facility's construction, in addition to its design. It is feasible to ascertain the services needed and pricing for the contractors while preparing the bid and also enables correct invoicing at a later point by providing the digital building model as part of the tendering process. A 4D Building Information Model may be used to assess the construction sequence, identify spatial collisions, and arrange site logistics by linking the various building components with the anticipated construction timeframes. In addition, a 5D model incorporates cost data and may be used to estimate how costs change over time [29]. Lastly, BIM techniques may also help with issue management and the billing of building projects.

The advantages of BIM during the building execution phase are as follows [36, 37]:

- Enhancing the capacity of contractors to make informed decisions through the estimation, coordination, and scheduling of the construction process;
- Improving the comprehension of the sequence and duration of construction activities;
- Improving the visualization of construction details;
- Improving the synchronization of design and construction planning;
- Improving constructability;
- Improving risk identification and the suitability of risk management to be made;
- Improving safety features;
- Improving productivity through time and cost savings;

- Reducing errors, waste, and rework for better sustainability for construction; Immediate response to design changes;
- Optimizing client experience and satisfaction;
- Increasing workforce effectiveness.

There are several advantages of applying BIM throughout the building phase that improve the general effectiveness, caliber, and safety of construction projects. Construction teams may save risks, cut down on errors, and enhance project results by using BIM for planning, coordination, cost management, procurement, quality control, and safety [37]. Modern construction technologies and BIM integration significantly simplify procedures and promote the timely completion of building projects.

3.3 Facilities operation and management phase

Building information modeling (BIM) is still useful for facilities operation and management (O&M) after construction is complete. BIM is an essential tool for effectively managing and preserving the built environment throughout this time [38].

The International Facility Management Association (IFMA) defines facility management as a multidisciplinary profession that integrates people, location, process, and technology to guarantee the built environment functions properly. Roles and duties in facility management are divided into eight main categories. Planning, budgeting, space management, interior design, interior installation, architecture, engineering services, building maintenance, and operations are some of the topics covered in this [38].

Building information modeling (BIM) is still useful for facilities operation and management (O&M) after construction is complete. BIM is an essential tool for effectively managing and preserving the built environment throughout this time [38]. Using the digital building model during the course of a constructed facility's rather lengthy operation period yields additional benefits to the BIM technique. The orderly transfer of BIM data from the design team to the owner, including all pertinent data from the building phase, is an essential precondition. The owner may input high-value digital data straight into his facility or asset management systems in place of "dead" drawings. This implies that, with regard to buildings, data regarding room dimensions, HVAC, energy, and telecommunications may be accessed instantly and does not require human entry [38, 39]. Information on installed devices, such as maintenance schedules and warranty terms, is very useful for the building's operation. Building information modeling (BIM) is still useful for facilities operation and management (O&M) after construction is complete. BIM is an essential tool for effectively managing and preserving the built environment throughout this time [38].

Maintaining the digital building model is crucial; any modifications made to the physical facility must be reflected in its digital counterpart. In the event that more extensive repairs or alterations are needed down the road, the building model serves as a great foundation for the essential design work. In order to arrange for ecologically responsible recycling or disposal of the materials used in the building of a constructed facility when its lifecycle comes to an end and it is set to be demolished, the digital twin offers comprehensive information about those materials [39].

The benefits of BIM in the phases of facilities, operations, and maintenance are as follows [37, 40]:

- Building information modeling (BIM) is very useful for facilities operation and management (O&M) after construction is complete.
- BIM is an essential tool for effectively managing and preserving the built environment throughout this time.
- It is easier to share information about a building's lifecycle.
- Collaboration is improved.
- Environmental sustainability is enhanced.
- Whole life cost control is improved.
- Emergency management is improved.
- Project closeout is enhanced.
- Project information loss is decreased.
- Built property is tracked.
- Proactive property management is done.
- Maintenance history reviews are enabled.

Building Information Modeling (BIM) is an excellent technique for managing constructed assets throughout their lifespan in the Facilities Operation and Management Phase. Facility managers can optimize operations, enhance sustainability, ensure regulatory compliance, and improve overall facility performance by utilizing Building Information Modeling (BIM) for as-built documentation, maintenance planning, energy management, space utilization, asset management, security planning, and financial analysis in their facilities. BIM integration with cutting-edge technology and data-driven tactics enhances effective facilities management and helps organizations accomplish their goals and objectives [39].

4. BIM and the construction industry

Although it still lags behind other industries in terms of productivity, efficiency, quality, and sustainability, the construction sector is one of the biggest in the world. This is due to its ineffectiveness and low output, which are attributed to the fragmented character of the project delivery process. When attempting to creatively incorporate construction into project development, the construction industry frequently encountered obstacles. These disadvantages frequently result in poor project performance and project delays [41]. Improving the industry's performance requires an awareness of the factors that contribute to these disadvantages. Thus, scholars from all around the world have looked into what causes some of the industry's problems. The examined literature reveals that a primary contributing factor is inadequate communication among project players [42]. A never-ending loop of information sharing between project stakeholders is required while doing construction projects, as noted in Ref. [43]. Both inadequate documentation and information management have contributed to the fragmentation of the construction industry's operations. As a result, a number of studies have been carried out to look at ways to enhance project delivery by reducing the issue of inadequate communication in the traditional techniques of project execution [6]. The use of Building Information Modeling is one of these.

With the development of information technology (IT), the globe is now referred to as a global village. No industry in the world is immune to the advantages of information technology. The built environment's cutting edge has been greatly aided by the advent of information technology in the construction sector. The building industry has developed a variety of software as a result of the integration of information technology (IT), and BIM is among the newest and most popular programs in the sector. It is indisputable that the use of BIM in the built environment has improved construction professionals' performance in terms of delivering infrastructures more quickly and better. The BIM system generates models that are extremely rich in data that can be analyzed by individual professionals in the construction sector to help them make decisions about the delivery of infrastructure, according to Associated General Contractors of America [44].

Because BIM enables construction professionals to work together as an integrated team during the project delivery process, Rokooei [45] further highlighted the significance of BIM in the industry. There has never been a synergy platform like the one that BIM offers to all parties involved in the building sector. However, the result of these platforms has been a geometric rise in the delivery of infrastructure production. The substantial potential of BIM for management, optimization, sustainability, and communication has been demonstrated by several research [46]. In actuality, collaborative BIM solutions facilitate improved lifecycle data management [47], claims management [46], site management [45], coordination and communication among stakeholders, and design correctness and installation quality.

Meanwhile, several studies [5, 48] have shown its capacity to envisage planned structures prior to the erection stage [49] in order to foresee all types of dangers, boost productivity, save money, and improve teamwork. In reality, Building SMART reveals [49] that BIM contributed to improvements in understanding design objectives, overall project quality, and cost management in Europe and North America, respectively, of 69, 62%, and 43, 65, 54, and 37%. However, BIM assisted in decreasing disputes among project participants, making modifications during construction, and providing clarification on requests by 59, 56, and 43% in Europe and 68, 54, and 47% in North America, respectively. Additionally, BIM enables significant cost savings, particularly in the AEC project lifecycle's construction and exploitation stages.

Cost estimate, collision detection, energy simulation, quantity take-off, schedule simulation, facility management, health and safety, off-site manufacturing, and three-dimensional representation are only a few of the current applications for BIM [46, 50]. The mid-2000s witnessed a variety of moves toward the global adoption of BIM in order to address low inefficiencies and other barriers to innovation in the Architecture, Engineering, and Construction (AEC) industry [51]. The adoption of BIM is being driven mostly by the increased public knowledge of it. But even with

BIM's global debut, its outputs are uneven, most likely as a result of acceptance and implementation issues in various nations.

5. BIM adoption around the world

Global adoption of Building Information Modeling (BIM) varies greatly depending on a number of factors, including industry standards, cultural norms, legal constraints, and technical maturity.

Instead of strategic adoption or use, research appears to have been concentrated on software innovations for the BIM process [52]. Although building projects in industrialized nations are gradually requiring BIM, poorer nations are trailing behind in the development of this technology. Both developed and developing nations still struggle with the adoption of BIM; but, due to the coordinated efforts of certain of their governments and organizations, developed nations have surpassed poor nations in this regard [53].

Table 1 provides a summary of the adoption of BIM across various geographies.

Similar to industrialized countries, the construction sector in emerging nations is composed of over 90% small and medium-sized firms (SMEs) [48], and it is often regarded as the engine of economic progress. Compared to large organizations, these smaller ones have an easier time making changes to their organizational structure. They also carry out smaller-scale initiatives, which facilitate innovation and may sometimes provide more advantages in a shorter amount of time than larger-scale projects [55]. However, because it involves a significant financial commitment and level of risk, SMEs frequently refuse to engage in innovations that are outside of their comfort zones. They are quick to embrace innovations that fit within the current organizational capabilities and may quickly benefit the business. However, the SMEs view the implementation of BIM as a radical process [48], disruptive [56], a significant shift [55], and challenging [57]. This is because it is believed that large companies with "organizational slack," which encourages risk-taking and innovative experimentation, can afford to invest in the initial costs of implementing BIM quickly, but SMEs cannot [58].

5.1 Factors influencing global BIM adoption

The adoption of Building Information Modeling (BIM) around the world is influenced by a variety of factors. These factors can either drive or hinder the implementation and integration of BIM in construction projects.

5.1.1 Government mandates and policies

In many countries, government mandates requiring the use of BIM for public sector projects have significantly boosted adoption. Examples include the UK's BIM Level 2 mandate and Singapore's government BIM requirements. Also, Government incentives, grants, and support programs can encourage private sector adoption by reducing the financial and technical barriers to implementing BIM.

5.1.2 Industry standards and initiatives

Industry bodies and professional organizations often develop BIM standards and guidelines, which help standardize practices and encourage widespread adoption.

Countries	BIM Espousal
United States	BIM adoption is widespread across the construction industry, driven by government mandates, industry initiatives (e.g., AIA's BIM documents), and the push for efficiency and sustainability. Major projects often require BIM implementation.
Canada	BIM adoption is also significant, with provincial governments and industry associations promoting its use. Large construction firms and government projects frequently use BIM for design, construction, and facilities management.
United Kingdom	The UK has been a leader in BIM adoption, driven by the government's BIM Level 2 mandate since 2016. Public sector projects require BIM for collaboration and lifecycle management.
Scandinavia	Countries like Sweden and Norway have embraced BIM for its efficiency and sustainability benefits. BIM is widely used in both public and private-sector construction projects.
Australia and New Zealand	BIM adoption is growing rapidly, supported by government initiatives and industry mandates. Major infrastructure projects and large-scale developments increasingly require BIM implementation.
Singapore	Known for its advanced construction industry, Singapore mandates BIM for government projects and promotes its use through incentives and training programs.
Saudi Arabia	BIM adoption is gaining traction as part of Vision 2030 initiatives to modernize the construction sector and improve project efficiency.
Brazil	BIM adoption is growing, supported by government policies to improve infrastructure and construction project efficiency. Major cities and public projects increasingly require BIM implementation.
Mexico	BIM adoption varies by region but is increasing, driven by large-scale infrastructure projects and efforts to modernize construction practices.
South Africa	BIM adoption is advancing, particularly in urban centers and major construction projects. The government promotes BIM as part of efforts to enhance infrastructure development and efficiency.
China	China is regarded as the most advanced in using BIM, especially by large enterprises within the AEC industry, as it aspires to be a major global player and transition into a developed nation. In China, the use of BIM has progressively moved from the design to the building stages. Nevertheless, the majority of BIM's use is currently restricted to visualizing conflicts in building systems. Dou et al. [54] came to the conclusion that BIM standards, regulations, and business models are essential to promoting greater BIM growth in China via case studies of thirty BIM projects.

Table 1.
BIM adoption among countries around the world [18, 28, 48, 51].

Collaborative efforts within the industry, such as consortiums and BIM forums, promote knowledge sharing and establish best practices.

5.1.3 Technological readiness

The availability and affordability of BIM software and tools are crucial. Regions with access to advanced technological infrastructure tend to have higher adoption rates. Also, compatibility and integration with other construction technologies and software systems (e.g., CAD, ERP) are essential for smooth BIM implementation.

5.1.4 Educational and training programs

The availability of BIM training programs and educational courses for professionals in the architecture, engineering, and construction (AEC) industry is vital for adoption. Incorporating BIM into university curricula helps prepare the next generation of professionals with the necessary skills and knowledge.

5.1.5 Project complexity and size

BIM is particularly beneficial for large-scale and complex projects where its capabilities in coordination, visualization, and data management can significantly enhance project outcomes. Increasingly, clients are specifying BIM as a requirement for project delivery, driving adoption among construction firms.

5.1.6 Cultural and organizational factors

The willingness of organizations to change traditional processes and adopt new technologies plays a significant role in BIM adoption. In some regions, cultural resistance to change and a preference for traditional methods can hinder BIM adoption.

5.1.7 Market demand and competition

Competitive pressure within the construction industry can drive BIM adoption as firms seek to gain a competitive edge through enhanced efficiency and innovation. As clients become more aware of the benefits of BIM, their demand for BIM-based project delivery increases, driving adoption across the industry.

5.1.8 Integration with other technologies

The integration of BIM with other technologies like the Internet of Things (IoT), augmented reality (AR), and virtual reality (VR) enhances its capabilities and promotes adoption. The use of BIM in conjunction with construction automation technologies, such as robotics and prefabrication, further drives its adoption.

5.1.9 Environmental and sustainability goals

BIM's ability to support sustainable design and construction practices makes it attractive in regions with stringent environmental regulations and sustainability goals. Furthermore, achieving green building certifications (e.g., LEED) often requires detailed environmental performance data, which BIM can provide.

5.2 Barriers to BIM adoption

Despite the numerous benefits of Building Information Modeling (BIM), its adoption in the construction industry faces several barriers. These obstacles can vary by region, project type, and organizational context.

5.2.1 High initial costs

The cost of purchasing BIM software and upgrading hardware to support it can be significant, especially for small and medium-sized enterprises (SMEs). Expenses related to training, hiring skilled personnel, and changing workflows can also be substantial.

5.2.2 Lack of skilled personnel

There is often a shortage of professionals who are trained and experienced in using BIM. This skills gap can delay adoption and reduce the effectiveness of BIM implementation. Resistance from employees who are accustomed to traditional methods and hesitant to adopt new technologies can also impede BIM adoption.

5.2.3 Fragmented industry practices

The construction industry is highly fragmented, with many stakeholders involved. Achieving the level of collaboration required for effective BIM use can be challenging. Furthermore, variability in standards and practices across different regions and projects can complicate BIM implementation.

5.2.4 Interoperability issues

Different stakeholders might use different BIM software tools, leading to compatibility issues and difficulties in data exchange. Integrating BIM with other existing systems and technologies (e.g., ERP, CAD) can be complex and problematic.

5.2.5 Limited awareness and understanding

BIM can be perceived as complex and difficult to implement, leading to a reluctance among potential adopters, and lack of awareness about the benefits and capabilities of BIM can result in its low adoption rates.

5.2.6 Legal and contractual issues

Issues related to intellectual property, data ownership, and liability for errors in the BIM model can be significant barriers, and traditional contractual frameworks may not support the collaborative nature of BIM, requiring new approaches to project delivery and management.

5.2.7 Limited client demand

If clients do not demand BIM in their project specifications, construction firms might not see the incentive to adopt it. In regions or sectors where market demand for BIM is low, there is less motivation for firms to invest in BIM.

5.2.8 Regulatory and policy barriers

In regions without government mandates or policies promoting BIM, adoption can be slow. Regulatory requirements and compliance issues can complicate BIM adoption, especially in highly regulated sectors. For example, in Indonesia, lack of expertise, high cost of implementation, and resistance to change are the major barriers hindering BIM adoption in SME firms [48].

6. Case studies

6.1 Sydney opera house, Australia

The Sydney Opera House, a UNESCO World Heritage site, underwent a major BIM implementation as part of its Facilities Management Exemplar Project. The objectives of the project were to improve asset management and maintenance processes, enhance the building's operational efficiency, and preserve the historical and architectural integrity of the structure [59, 60].

The following were implemented with BIM:

- Creation of a detailed BIM model that included 3D geometry and metadata for all components.
- Integration of the BIM model with existing facilities management software.
- Use of laser scanning and photogrammetry to capture accurate as-built conditions.

With the use of BIM, the Sydney Opera House was able to ensure:

- Enhanced accuracy in asset data management, reducing maintenance errors.
- Improved efficiency in scheduling and conducting maintenance tasks.
- Better visualization tools for planning and executing preservation work.

6.2 The Cathay Pacific cargo terminal, Hong Kong

The project, which is estimated to have cost USD $500 million, is situated on the south side of the Hong Kong International Airport platform, home to the cargo terminal facilities for Cathay Pacific (CX). At 2.6 million tons of cargo annually throughput, the facility will be the largest air cargo terminal in the world (measured in tons per square meter). During peak hours, the cargo terminal handles about 75 flights daily, carrying up to 110 tons of cargo from modified 747 cargo carriers known as silver bullets, or up to 25 tons from passenger planes.

The client stipulated that the BIM method must be used to coordinate the design from the start of the design phase. CX was committed to using cutting-edge technologies to lower the possibility of delays and cost overruns on the intricate fast-track project. The cargo facility is a specially designed structure that needs sophisticated mechanical, electrical, ventilation, drainage, and specialized mechanical systems. A thorough clash detection analysis matrix can be used to find and resolve any design conflicts between different systems (cargo handling, structural, architectural, municipal systems, and specialized systems) as the design, drawings, and specifications for these systems are integrated into the BIM.

Conflicts of many kinds need to be recognized in a project like this one. When plumbing or other systems cross through or otherwise obstruct structural or architectural features, there will be a severe collision. If this is disregarded during the design phase, it may result in a delay on site and need the filing of an RFI.

Pallets weighing six tons each are used to bring cargo inside the building, where it is handled by the Materials Handling System (MHS). The MHS is made up of a semi-automated roller-deck assembly that moves the cargo pallets laterally. The pallets are raised or lowered vertically between floors using big hoists. This means that as cargo pallets are transported and processed across the facility, the BIM must take into consideration the many systems that can have a clearance dispute with their "kinetic envelope."

A unique clash matrix was used to identify conflicts, and it included operating needs and design specifications (headroom, MHS standards, structural and architectural coordination, and MEP coordination) in the study (**Figure 3**). Priorities are set for the analysis, and problems are fixed before any building ever starts. Throughout the design coordination phase, this is an ongoing procedure that is worked on by a committed engineer. As a result, there are far fewer on-site RFIs, delays, and cost overruns.

The project has operated more smoothly as a consequence of CX's proactive strategy, which also made previously divided design process components—structural engineers, architects, MEP, etc.—cooperate methodically from the beginning. This has necessitated comprehensive retraining for key staff in the usage of standardized software platforms and systems to guarantee uniformity in data design and presentation.

"At times we have had to insist that sliding back to 2-D methods is just not acceptable" is the professional inertia that is causing resistance. The strategy appeared to be changing the industry's culture in the way that was intended (**Figure 5**).

These case studies highlight the diverse applications and benefits of BIM in the construction industry, demonstrating how it can enhance project coordination, reduce errors, improve efficiency, and support sustainability goals [62].

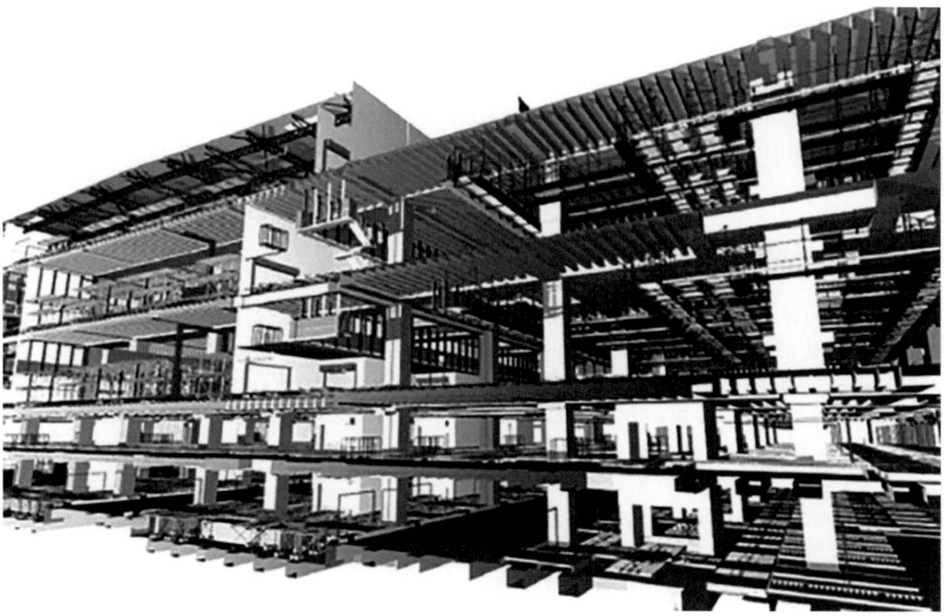

Figure 5.
Scheme of the new CX cargo terminal of Hong Kong [61].

7. Conclusion

Building Information Modeling (BIM) has become a pivotal technology in transforming the construction industry, offering a multitude of benefits that enhance efficiency, collaboration, and sustainability. This review has explored the implementation and practices of BIM, drawing several key conclusions:

7.1 Enhanced collaboration and communication

BIM significantly improves collaboration and communication among stakeholders by providing a unified digital platform. This fosters better coordination, reduces errors, and enhances decision-making throughout the project lifecycle.

7.2 Increased project efficiency and cost savings

The use of BIM leads to greater project efficiency and cost savings. Its ability to produce accurate 3D models helps in the early detection of potential issues, reducing the likelihood of costly changes and delays. BIM also facilitates better resource planning and allocation, optimizing project workflows and timelines.

7.3 Sustainability and lifecycle management

BIM supports sustainable construction practices by enabling comprehensive energy analysis and material selection. It extends its utility beyond the construction phase into facilities management, providing valuable data for ongoing maintenance and operations, thereby enhancing the lifecycle management of buildings.

7.4 Integration with emerging technologies

The integration of BIM with technologies such as IoT, AR, and AI enhances its functionality. These technologies enable advanced data analytics, predictive maintenance, and improved visualization, driving innovation and improving project outcomes.

7.5 Challenges and barriers to adoption

Despite its advantages, BIM adoption faces several barriers. High initial costs, lack of skilled professionals, interoperability issues, and organizational resistance are significant challenges. Overcoming these requires concerted efforts from all industry stakeholders, including targeted training, supportive policies, and increased awareness of BIM's benefits.

7.6 Global variability in adoption

BIM adoption varies significantly across the globe, influenced by regional factors such as government mandates, industry standards, and technological readiness. While some countries have seen widespread BIM implementation, others lag due to limited resources and regulatory frameworks.

Acknowledgements

The author expresses gratitude for the assistance received from Kampala International University.

Conflict of interest

The author declares that there is no conflict of interest.

Author details

Bamidele Charles Olaiya[1*], Olaolu George Fadugba[2]
and Mustapha Muhammad Lawan[1]

1 Department of Civil Engineering, School of Engineering and Applied Sciences (SEAS), Kampala International University, Ishaka, Uganda

2 Department of Civil Engineering, Federal University of Texhnology, Akure, Nigeria

*Address all correspondence to: bmolaiya@kiu.ac.ug

References

[1] Eastman C, Teicholz P, Sacks R, Liston K. BIM Handbook: A Guide to Building Information Modeling for Owners, Managers, Designers, Engineers, and Contractors. 2nd ed. Hoboken, NJ: Wiley; 2008. DOI: 10.1002/9780470261309

[2] Aguome NM, Alaneme GU, Olaiya BC, Lawan MM. Evaluation of lean construction practices for improving construction project delivery. Case study of Bushenyi District, Uganda. Cogent Engineering. (Abingdon, UK: Taylor & Francis). 2024;**11**(1):1-15. DOI: 10.1080/23311916.2024.2365902

[3] Azhar S. Building information modeling (BIM): Trends, benefits, risks, and challenges for the AEC industry. Leadership and Management in Engineering. 2011;**11**(3):241-252

[4] Olaiya BC, Lawan MM, Olonade KA. Utilization of sawdust composites in construction—A review. SN Applied Sciences. 2023;**5**:140. DOI: 10.1007/s42452-023-05361-4

[5] Ghaffarianhoseini A, Tookey J, Ghaffarianhoseini A, Naismith N, Azhar S, Efimova O, et al. Building information modelling (BIM) uptake: Clear benefits, understanding its implementation, risks and challenges. Renewable and Sustainable Energy Reviews. 2017;**75**:1046-1053

[6] Khosrowshahi F, Arayici Y. Roadmap for implementation of BIM in the UK construction industry. Engineering, Construction and Architectural Management. 2012;**19**(6):610-635

[7] Succar B. Building information modelling framework: A research and delivery Foundation for Industry Stakeholders. Automation in Construction. 2009;**18**(3):357-375

[8] Succar B, Sher W, Williams A. Measuring BIM performance: Five metrics. Architectural Engineering and Design Management. 2012;**8**:120-142. DOI: 10.1080/17452007.2012.659506

[9] Miettinen R, Paavola S. Beyond the BIM utopia: Approaches to the development and implementation of building information modeling. Automation in Construction. 2014;**43**:84-91. DOI: 10.1016/j.autcon.2014.03.009

[10] Coates P, Arayici Y, Koskela L, Kagioglou M, Usher C, O'Reilly K. The limitations of BIM in the architectural process. In: ICSU 2010: First International Conference on Sustainable Urbanism, 15-17th December 2010. Hung Hom, Kowloon, Hong Kong: Harbour Grand Kowloon; 2010

[11] Kunz J, Fischer M. Virtual design and construction: Themes, case studies and implementation suggestions. CIFE Working Paper. 2012;**97**:50. Available from: https://purl.stanford.edu/gg301vb3551

[12] Austin R, Devin L. Artful making: What managers need to know about how artists work. Prentice Hall. Upper Saddle River, NJ: Prentice Hall; 2003. ISBN 0-13-008695-9

[13] Ford R, Piccolo R, Ford L. Strategies for building effective virtual teams: Trust is key. Business Horizons. (Amsterdam, Netherlands: Elsevier). 2016;**60**(5):567-576. DOI: 10.1016/j.bushor.2016.08.009

[14] Nytsch-Geusen C. Building simulation on a large scale; components,

spaces, buildings, city districts. Bauphysik. 2013;**35**:1-1

[15] Kiani I, Nobahar A, Khalili GS, Marsono A. Barriers to implementation of building information modeling in scheduling and planning phase in Iran. Journal of Building Performance. 2015;**9**(5):91-97

[16] Taylor J, Levitt R. Inter-organizational knowledge flow and innovation diffusion in project-based industries. In: Proceedings of the 38th Annual Hawaii International Conference on System Sciences (HICSS). IEEE Computer Society. 2005. pp. 1-10. DOI: 10.1109/HICSS.2005.334

[17] Deng ZM, Li H, Tam CM, Shen QP, Love PED. An application of the internet-based project management system. Automation in Construction. 2001;**10**:239-246

[18] Gu N, London K. Understanding and facilitating BIM adoption in the AEC industry. Automation in Construction. 2010;**19**(8):988-999

[19] Pillay N, Musonda I, Makabate C. Use of BIM at higher learning institutions: Evaluating the level of implementation and development of BIM at built environment schools in South Africa. In: Proceedings of the 2018 International Conference on Construction and Real Estate Management. IOP Publishing; 2018

[20] Barlish K, Sullivan K. How to measure the benefits of BIM—A case study approach. Automation in Construction. 2012;**24**:149-159. DOI: 10.1016/j.autcon.2012.02.008

[21] Ahmad LA, Mohd S, Rakiman U. Potential improvement of building information modeling (BIM). Implementation in Malaysian Construction Projects. 2015. DOI: 10.13140/RG.2.1.4661.8642

[22] Pinti L, Ricardo C, Serena B. A review of building information modelling (BIM) for facility management (FM): Implementation in public organisations. Applied Sciences. 2022;**12**(3):1540. DOI: 10.3390/app12031540

[23] Cao Y, Syahrul NK, Mardhiyah NA. Building information modeling (BIM) capabilities in the operation and maintenance phase of green buildings: A systematic review. Buildings. 2022;**12**(6):830. DOI: 10.3390/buildings12060830

[24] Ekholm A. Activity objects in CAD programs for building design. In: Proceedings of CAAD Futures 2001. Eindhoven, Netherlands: Technische Universiteit Eindhoven (TUE); 2001. pp. 61-69

[25] Latiffi AA, Brahim J, Fathi MS. The development of building information modeling (BIM) definition. Applied Mechanics and Materials. 2014;**567**:625-630

[26] Alhusban M, Al-Bizri S, Danso-Amoako M, Gaterell M. Procurement route and building information modelling (BIM) implementation effect on sustainable higher education refurbishment projects. In: Advances in Human Factors, Business Management and Leadership. Cham, Switzerland: Springer; 2017. DOI: 10.1007/978-3-319-50346-2_8

[27] Abanda H, Vidalakis C, Oti A, Tah J. A critical analysis of building information modelling systems used in construction projects. Advances in Engineering Software. 2015;**90**:183-201. DOI: 10.1016/j.advengsoft.2015.08.009

[28] Rogers J, Chong H, Preece C. Adoption of building information modelling technology (BIM): Perspectives from Malaysian engineering consulting services firms. Engineering, Construction and Architectural Management. 2015;**22**:424-445. DOI: 10.1108/ECAM-05-2014-0067

[29] Hasan A, Rasheed S. The benefits of and challenges to implement 5D BIM in construction industry. Civil Engineering Journal. 2019;**5**:412. DOI: 10.28991/cej-2019-03091255

[30] Onungwa I, Uduma-Olugu N. Building information modelling and collaboration in the Nigerian construction industry. Journal of Construction Business and Management. 2017;**1**:1-10. DOI: 10.15641/jcbm.1.2.53

[31] Kamardeen I. 8D BIM modelling tool for accident prevention through design. In: Egbu CO, Lou EC, editors. 26th Annual Conference-ARCOM. Reading: Association of Researchers in Construction Management; 2010. pp. 281-289

[32] Jadhav PP. Assessment of building energy by performing simulation with BIM. International Journal of Emerging Technologies and Innovative Research. Jun 2019;**6**(6):393-399. ISSN:2349-5162. Available from: http://www.jetir.org/papers/JETIR1906P60.pdf

[33] Ershadi M, Jefferies M, Davis P, Mojtahedi M. Implementation of building information modelling in infrastructure construction projects: A study of dimensions and strategies. International Journal of Information Systems and Project Management. Project Management Institute. 2021;**9**(4):43-59. DOI: 10.12821/ijispm090403

[34] Kiviniemi A. Challenges of Interoperable BIM in a between Organization. 2010. Available from: http://aarch.dk/fileadmin/filer/Sune/Arto_Kiviniemi.pdf

[35] Koskela LJ. Making do - the eighth category of waste. In: 12th Annual Conference of the International Group for Lean Construction, 3-5 August 2004. Helsingborg, Denmark; 2004

[36] Love P. In search of the magic bullet: Building informational modelling, garbage in, gospel out. In: Working Paper, Curtin University. 2010

[37] McGraw-Hill Construction. The business value of BIM for construction in major global markets: How contractors around the world are driving innovation with building information modeling. Smart Market Report. New York, NY: McGraw-Hill Construction; 2014

[38] Arayici Y, Aouad G. Building information modelling (BIM) for construction lifecycle management. In: Construction and Building: Design, Materials, and Techniques. NY, USA: Nova Science Publishers; 2010. pp. 99-118

[39] Arayici Y, Coates P, Koskela LJ, Kagioglou M, Usher C, O'Reilly, K. Technology adoption in the BIM implementation for lean architectural practice. Automation in Construction. 2011;**20**(2):189-195

[40] Arayici Y, Coates P, Koskela LJ, Kagioglou M, Usher C, Reilly K. BIM adoption and implementation for architectural practices. Structural Survey. 2011;**29**(1):7-25

[41] Saraireh I, Haron A. Understanding the conceptual of building information modeling: A literature review. International Journal of Civil Engineering and Technology (IJCIET). 2020. DOI: 10.34218/ijciet.11.1.2020.018

[42] PMI. A Guide to the Project Management Book of Knowledge (PMBOK). 4th ed. Newtown Square: Project Management Institute; 2008

[43] Ku K, Mojtaba MS. BIM experiences and expectations: The constructors' perspective. International Journal of Construction Education and Research. 2011;**7**:175-197. DOI: 10.1080/15578771.2010.544155

[44] Associated General Contractors of America. The Contractor's Guide to BIM. Las Vegas: AGC Research Foundation; 2005

[45] Rokooei S. Building information modeling in Project Management: Necessities, challenges and outcomes. Procedia - Social and Behavioral Sciences. 2015;**210**:87-95. DOI: 10.1016/j.sbspro.2015.11.332

[46] Gerrish T, Ruikar K, Cook M, Johnson M, Phillip M, Lowry C. BIM application to building energy performance visualization and management: Challenges and potential. Energy and Buildings. (Amsterdam, Netherlands: Elsevier). 2017;**144**:162-173. DOI: 10.1016/j.enbuild.2017.03.032

[47] Saka A, Chan D. A global taxonomic review and analysis of the development of BIM research between 2006 and 2017. Construction Innovation. (Bingley, UK: Emerald Publishing). 2019;**19**(1):26-51. DOI: 10.1108/CI-12-2018-0097

[48] Saka A, Chan D. Adoption and implementation of building information modelling (BIM) in small and medium-sized enterprises (SMEs): A review and conceptualization. Engineering Construction & Architectural Management. 2020. DOI: 10.1108/ECAM-06-2019-0332

[49] Boukara A, Aziz N. A brief introduction to building information modeling (BIM) and its interoperability with TRNSYS. Renewable Energy and Sustainable Development. 2016;**1**(1):126-130. DOI: 10.21622/resd.2015.01.1.126

[50] Aires M, López-Alonso M, Martinez-Rojas M. Building information modeling and safety management: A systematic review. Safety Science. 2018;**101**:11-18. DOI: 10.1016/j.ssci.2017.08.015

[51] Mehran D. Exploring the adoption of BIM in the UAE construction industry for AEC firms. Procedia Engineering. 2016;**145**:1110-1118. DOI: 10.1016/j.proeng.2016.04.144

[52] Hjelseth E. BIM understanding and activities. WIT Transactions on the Built Environment. (Southampton, UK: WIT Press). 2017;**170**:3-14. DOI: 10.2495/BIM170011

[53] Olawumi T, Chan D. Development of a benchmarking model for BIM implementation in developing countries. Benchmarking: An International Journal. (Bingley, UK: Emerald Publishing). 2019;**26**(8):2294-2315. DOI: 10.1108/BIJ-05-2018-0138

[54] Dou Y, Bo Q. Characteristics and dynamics of BIM adoption in China: Social network analysis. Journal of Construction Engineering and Management. (Reston, VA: American Society of Civil Engineers (ASCE)). 2022;**148**(3):04022007. DOI: 10.1061/(ASCE)CO.1943-7862.0002276

[55] Arayici Y, Kiviniemi A, Coates S, Koskela L, Kagioglou M, Usher C, et al. BIM implementation and adoption process for an architectural practice. In: Proceedings of the 29th International Conference on Education and Research in Computer Aided Architectural Design in Europe (eCAADe). Newcastle upon Tyne, UK: Association for Computer Aided Architectural Design in Europe; 2011

[56] Hochscheid E, Halin G. A model to approach BIM adoption process and possible BIM implementation failures. Creative Construction Conference 2018. Ljubljana, Slovenia; Jun 2018. pp. 257-264. ff10.3311/CCC2018-034ff. ffhal-02549314f

[57] Hosseini MR, Maghrebi M, Akbar NA, Martek I, Arashpour M. Analysis of citation networks in building information modeling research. Journal of Construction Engineering and Management. (Reston, VA: American Society of Civil Engineers (ASCE)). 2018;**144**(11):04018117. DOI: 10.1061/(ASCE)CO.1943-7862.0001492

[58] Kori S, Itanola M, Saka A. The capability and support of structural capital on BIM innovation in SMEs. International Journal of Knowledge Management. IISTE. 2019;**9**(2):56-66. DOI: 10.7176/IKM/9-2-06

[59] Murphy ME, McGovern E, Pavia S. Historic building information modelling (HBIM). Structural Survey. 2013;**31**(4):267-283. DOI: 10.1108/SS-01-2013-0004

[60] Yung P, Yip B. Construction quality in China: Innovative case study of the Shanghai world financial center project. Journal of Construction Engineering and Management. 2010;**136**(8):927-932. DOI: 10.1061/(ASCE)CO.1943-7862.0000205

[61] Rowlinson S, Collins R, Tuuli M, Jia A. Implementation of building information modeling (BIM) in construction: A comparative case study. AIP Conference Proceedings. 2010;**1233**:572-577. DOI: 10.1063/1.3452236

[62] Ghosh S, Rata B. One world trade center: Project Management innovation in the World's Most complex skyscraper. Proceedings of the Institution of Civil Engineers - Civil Engineering. 2012;**165**(6):56-62. DOI: 10.1680/cien.11.00052

Chapter 3

Dynamic Response Analysis of Coastal Bridge Members Exposed to Water Forces and Earthquakes

Riyadh Alsultani, Ibtisam R. Karim, Saleh I. Khassaf and Ahmed Ashor Al-Saadi

Abstract

Reinforced concrete pile foundations are a common and efficient solution for constructing deepwater bridges that span large bodies of water, such as rivers, seas, or portions of oceans. Many of these bridges are situated in areas with significant seismic hazards. These structures are not only subjected to typical water loads, such as currents and waves but also to earthquake forces. The interaction between these water forces and the structure generates hydrodynamic forces on the submerged parts of the bridges. Thus, accurately estimating these hydrodynamic forces during earthquakes is crucial for ensuring the structural safety of deepwater bridges. This chapter aims to assess the structural response of pile foundation bridge piers when exposed to hydrodynamic forces during earthquakes, utilizing DIANA FE software and parallel computation technology. The study model incorporates the combined effects of currents, waves, and earthquakes, along with the nonlinear behavior of soil and concrete. Using Stokes's fifth-order wave theory and Morison's hydrodynamic pressure formula, the wave force was applied as a distributed load on the bridge's pile foundation. The dynamic excitation characteristics of the pier under elastic conditions were analyzed, considering the influence of currents and waves.

Keywords: pile foundation, bridge pier, dynamic response, Morison's formula, hydrodynamic pressure, current-wave-earthquake

1. Introduction

The growth of the social economy has led to an increased demand for urban transportation, prompting the construction of new broad-span bridges across rivers and canals. These bridge piers, often situated in deep water, must withstand challenging environmental pressures [1]. Completed projects have shown that deepwater pile foundations can meet safety and performance criteria, passing dynamic effect tests related to wind, dampness, and flow. However, their seismic stability, particularly under the combined influence of significant ground motion and wave movement during a massive earthquake, remains to be fully confirmed [2–4].

Deepwater pile foundations, located several meters below the sea surface, are susceptible to hydrodynamic pressure from water waves and currents, as well as vibration induced by ground motion. The additional moment on the pile results from hydrodynamic pressure due to water waves and currents, affecting the internal forces within the pile [5]. Research on the dynamic response of water structures under seismic and wave action has provided useful insights. For example, Yamada et al. [6] used random vibration to study an offshore structure's dynamic response, representing waves with a Bretschneider power spectrum and ground motion with a Kanai power spectrum. Karadeniz [7] employed spectrum analysis to examine a three-dimensional structure's response under deepwater wave and seismic motion, treating both as stochastic processes. Fukusumi et al. [8] simulated waves as harmonic and analyzed dynamic effects influenced by structural stiffness, flow, and fluid density.

Etemad et al. [9] considered pile-soil interaction using the Nogami model to analyze seismic reactions of pile foundations under two scenarios: wave and ground motion in the same and opposite directions. Abbasi and Gharabaghi [10] incorporated nonlinearity into a 3D model of an offshore platform to analyze wave direction's impact on seismic response. Li et al. [11] used the additional mass method to create a pile-soil-pier-water model, examining dynamic water pressure and water level changes. He and Li [12] applied Morison's formula and wave theory to investigate seismic responses under combined earthquake and wave actions. Song et al. [13] addressed the issue of seismic response by developing a simplified calculation method for hydrodynamic pressures exerted on slender structures during earthquakes. Their research provides a valuable tool for engineers to efficiently estimate these pressures, which is essential for the design and safety assessment of bridges exposed to seismic activity. This method facilitates the prediction and management of dynamic responses, ensuring that structures can withstand the combined impacts of seismic and hydrodynamic forces.

Furthermore, the seismic hazards specific to certain regions underscore the importance of such studies. Al-Taie and Albusoda [14] examined the earthquake hazard in Iraq, using the Halabjah earthquake as a case study. Their research highlighted the seismic risks prevalent in Iraqi soil and the implications for infrastructure resilience. Understanding these hazards is crucial for designing bridges and other critical structures to endure and perform under earthquake conditions. Together, these studies contribute to a deeper understanding of the challenges posed by seismic activities on slender structures and provide practical solutions to enhance the resilience of infrastructure in earthquake-prone areas.

Recent studies have highlighted various aspects of these interactions. For instance, Li et al. [5] conducted underwater shaking table tests to evaluate the performance of a sea-crossing cable-stayed bridge under the combined action of earthquakes and waves. This research underscores the complex nature of multi-hazard scenarios and their cumulative effects on bridge stability. Similarly, Wei et al. [15] focused on the impacts of wave and wave-current actions on bridge towers, providing valuable insights into the hydrodynamic forces and their implications for structural design. The dynamic behavior of bridge piers under extreme conditions, such as dam-break floods, was explored by Huang and Liu [1]. Their analysis revealed how different directions of impact influence the response of bridge piers, which is crucial for designing resilient structures against such catastrophic events. Another study by Huynh et al. [16] examined the time-dependent seismic fragility of offshore bridges considering scour and chloride ion corrosion, highlighting the importance of accounting for long-term environmental effects in seismic assessments.

Incorporating the most recent studies into the literature review is essential for staying at the forefront of research in the field of deepwater pile foundations and bridge stability. As new methodologies, technologies, and theoretical advancements emerge, they offer novel insights and solutions that challenge and refine existing models and practices. For instance, recent research by Chen et al. [17] introduces an advanced numerical model for analyzing the effects of combined wave and seismic actions on deepwater pile foundations, providing more accurate predictions of structural responses compared to earlier approaches. Similarly, Zhao et al. [18] investigated the dynamic response of bridge piers under complex multi-hazard scenarios, highlighting the limitations of traditional models and proposing new strategies for enhancing structural resilience. By integrating these contemporary studies, the literature review not only reflects the latest developments but also identifies ongoing challenges and opportunities for future research. This dynamic and updated approach ensures that the research remains relevant and continues to advance the field by addressing contemporary issues and leveraging the latest innovations [19, 20].

Xu and Cai [21] conducted numerical simulations to explore how lateral restraining stiffness affects the dynamic response of bridge decks to solitary wave impacts. Their research offers a valuable framework for enhancing bridge deck designs to better withstand wave forces. Additionally, Qu et al. [22] evaluated the vulnerability of coastal bridges in the New York City metropolitan area by examining Hurricane Sandy as a case study. Their findings underscore the importance of robust design and adaptive strategies to safeguard coastal infrastructure against future extreme storm surges and waves.

Despite these studies, two main issues persist: the oversimplification of structures that neglect pile-soil interaction and the inadequate or ignored fluid influence modeling. Updating the literature review to include the most recent studies would ensure that the research remains at the cutting edge, reflecting the latest advancements and discussions in the field. This is essential for maintaining the chapter's relevance and authority. To accurately simulate the pier-piles-soil-water current-and-wave system under earthquake conditions, a 3D finite element model must be developed. This model should analyze bridge piers' dynamic responses, including relative displacement, acceleration, shear force, moment responses, and hydrodynamic pressure coefficients.

2. Case study description

This case study examines the soil-pile foundation-pier system of a multi-span bridge crossing the Songhua River in northeast China [23]. The bridge pier has a high rectangular cross-section, measuring 4.8 meters in length, 2.4 meters in width, and 15 meters in height. The pile cap features a square cross-section, standing 3 meters tall and 12 meters wide, supported by nine circular piles. These piles are 58 meters in height, with 12 meters extending above the scour line, a diameter of 1.8 meters, and a spacing of 4.5 meters.

The bridge deck, designed for two-way traffic, spans four lanes and measures 12 meters in width, with a top mass of 7.8 x 10^5 kg. The deck and pier are constructed from C35 concrete, which has Young's modulus of 31.5 GPa. In contrast, the pile cap and piles use C30 concrete, with Young's modulus of 28 GPa (Ministry of Communications of China [24]). **Figure 1** provides a sketch of the bridge pier with its elevated pile cap support.

The soil stratification overlying the bedrock is depicted according to [25]. **Table 1** lists the soil model parameters used in this study.

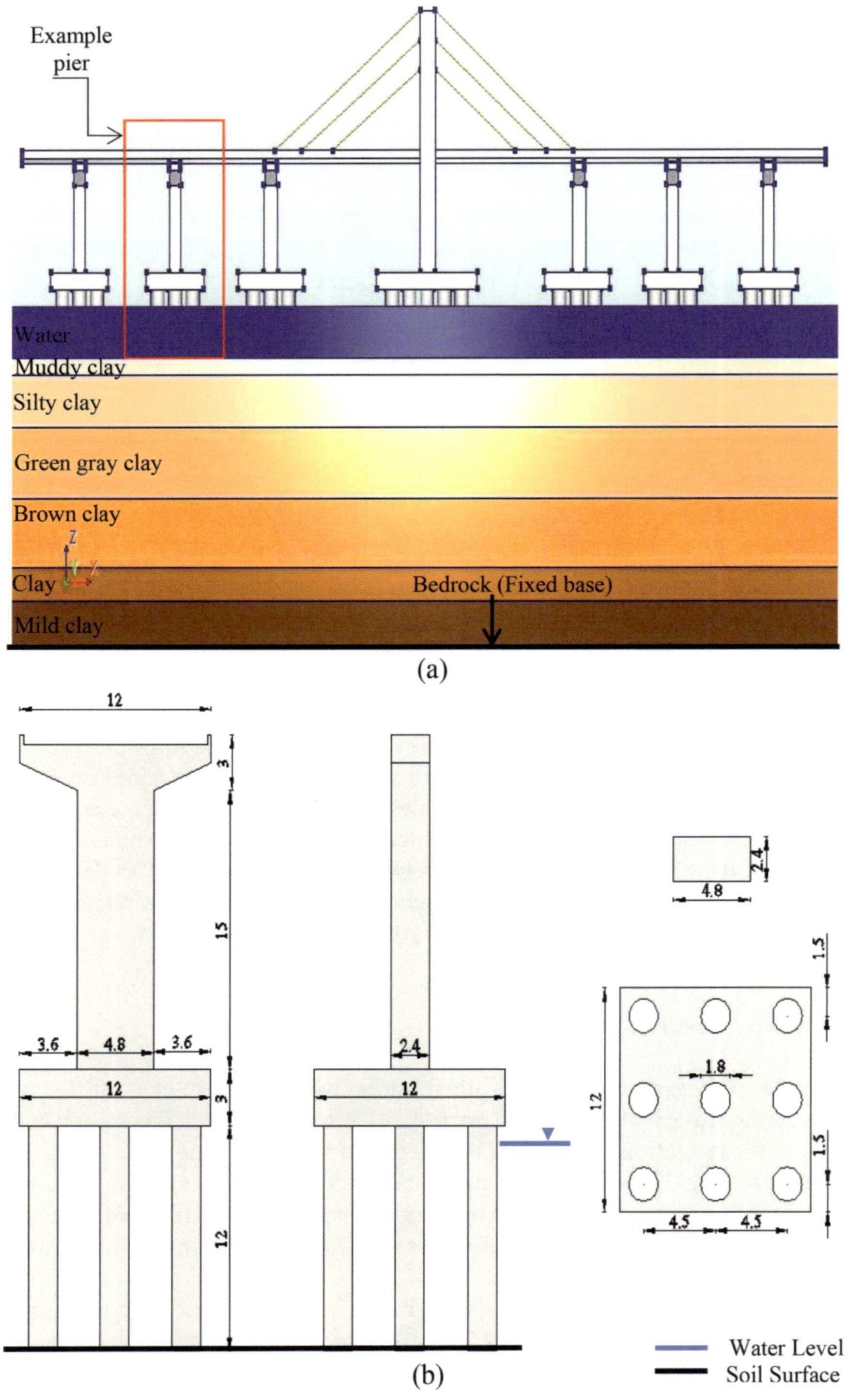

Figure 1.
Drawing of the selected case study: (a) geometry of the bridge; and (b) dimensions and details of the bridge members. All dimensions are in meters.

Soil layers	$\gamma_0\left(\times 10^{-4}\right)$	Density (Kg/m3)	Shear wave velocity (m/s)	Friction angle (°)
Mud	4.0	1800	170	16
Silt	4.0	1890	190	16
Green gray	3.7	1900	210	24
Brown clay	3.7	1960	260	24
Clay	3.8	1970	320	21
Mild clay	4.4	2030	380	21

Table 1.
Model parameters of soils [25].

According to the study, the border effects on the dynamic response of structures are sufficiently eliminated in this article because the foundation border is sufficiently big to do so. Concrete is used for the dick, pier, cap, and piles, and the plastic-damage model of concrete is used to describe the nonlinear behavior of concrete [26].

3. Combined current-wave and earthquake action

The water-bridge interaction that occurs as a result of an earthquake is taken into account in the examination of the dynamic response of bridge structures. The water's ability to travel is affected by the deformation or movement of these components, and the water's forces acting on the bridge take the form of hydrodynamic pressure. In a matrix form, the following represents the governing equation of transient structural dynamics (Morison et al., 1950):

$$[M]\left\{\ddot{x}(t)\right\}+[C]\left\{\dot{x}(t)\right\}+[K]\left\{x(t)\right\}=\left\{F_H(t)\right\}+[M]\left\{\ddot{x}_g(t)\right\} \tag{1}$$

Where [M], [C], and [K] represent the structural mass matrix, damping matrix, and stiffness matrix, respectively. $\{x(t)\}$, $\{\dot{x}(t)\}$, and $\{\ddot{x}(t)\}$ represent the structural relative displacement, velocity, and acceleration vectors, respectively. $\{x_g(t)\}$ is the acceleration vector of seismic ground motion. $\{F_H(t)\}$ is the fluid force vectors exerted on the bridge structure, including current-wave and earthquake-induced hydrodynamic forces.

The dynamic equation of motion can be represented as:

$$[M]\{\ddot{x}(t)\}+[C]\{\dot{x}(t)\}+[K]\{x(t)\}=C_M\rho V\left(\ddot{u}-\ddot{x}_0\right)+\frac{1}{2}C_D\rho A\left|\dot{u}-\dot{x}_0\right|\left(\dot{u}-\dot{x}\right)+\rho V\left\{\ddot{x}_g(t)\right\} \tag{2}$$

Where ρ is the water density; C_M and C_D are the coefficients of inertial and drag forces, respectively; V and A is the exposed face volume and area, respectively; x_0 and $\ddot{x}_0$ are the structure absolute velocity and structure associated acceleration, respectively; and $\dot{u}$ and $\ddot{u}$ are the water velocity and water-associated acceleration, respectively.

The hydrodynamic force per unit height, P, from Eq. (2) can be presented as shown in Eq. (3):

$$P = C_M \rho A\left(\ddot{u} - \ddot{x}_0\right) + \frac{1}{2} C_D \rho B \left|\dot{u} - \dot{x}_0\right|\left(\dot{u} - \dot{x}_0\right) + \rho A\left\{\ddot{x}_g(t)\right\} \tag{3}$$

Where B is the width of the exposed face.

The first term at the right of Eq. (3) denotes the inertia force, the second term denotes the drag force, and the third term denotes structural effect.

For coastal and offshore structures, the influence of hydrodynamic drag force, represented by combined current and wave actions, on the dynamic response of structures is very important [13, 27, 28]. Therefore, the Morison equation can be modified to account for currents by replacing '$\dot{u}$' by '$C + W$', where C is the water current speed and W is wave properties and as in Eq. (4):

$$P = C_M \rho A\left(\ddot{u} - \ddot{x}_0\right) + \frac{1}{2} C_D \rho B \left|(C + W) - \dot{x}_0\right|\left((C + W) - \dot{x}_0\right) + \rho A \ddot{u} \tag{4}$$

The inertial coefficient is calculated by simplifying the inertia force presented in the first part of the equation (Eq. 4) as the following expressions (Eq. 5):

$$C_M = \frac{M}{\rho H \frac{\pi}{4} D^2 \left(\ddot{u} - \ddot{x}_0\right)} \tag{5}$$

Where H is water depth and D is the diameter of the pile. As shown in Eq. (5), the hydrodynamic force is proportional to H and D^2, and independent of the action of frequency and amplitude.

The drag coefficient is calculated by simplifying the drag force presented in second part of equation (Eq. 4) as the following expressions (Eq. 6):

$$C_D = \frac{M}{\frac{1}{2} \rho H D \left|(C + W) - \dot{x}_0\right|\left((C + W) - \dot{x}_0\right)} \tag{6}$$

As shown in Eq. (6), the hydrodynamic drag coefficient is proportional to H and D, and independent of the current and wave.

A lift force is also connected to the loading on a bridge's submerged components, in addition to the drag force. Due to the orbital motion of the water particles, this lift force is perpendicular to the velocity vector and rotates around the axis of the members. The influence of the vortex shedding will be visible as noise in the measurements of the drag and inertia components since the magnitude, direction, and period of the lift force are unknown and cannot be applied to Morison's equation [29–31].

4. Computational model development

A soil-piles-pier model, illustrated in **Figure 2**, was created using parallel computation technologies and the DIANA software. The model employs a dynamic plastic-damage approach to accurately represent the dynamic behavior of concrete bridge components, modeling them as an assembly of beam elements. For the soil, a viscous-plastic memorial nested yield surface model was utilized to capture its dynamic nonlinearity [32–34]. The soil is modeled as an assembly of 8-node solid elements, assumed to be saturated and undrained during an earthquake, with a Poisson's ratio of 0.49.

The foundation soil model covers an area of 1000 meters by 1000 meters by 50 meters. To address the effects of region truncation on wave propagation, a 3D time-domain viscoelastic artificial boundary was applied. The soil layer corresponding to

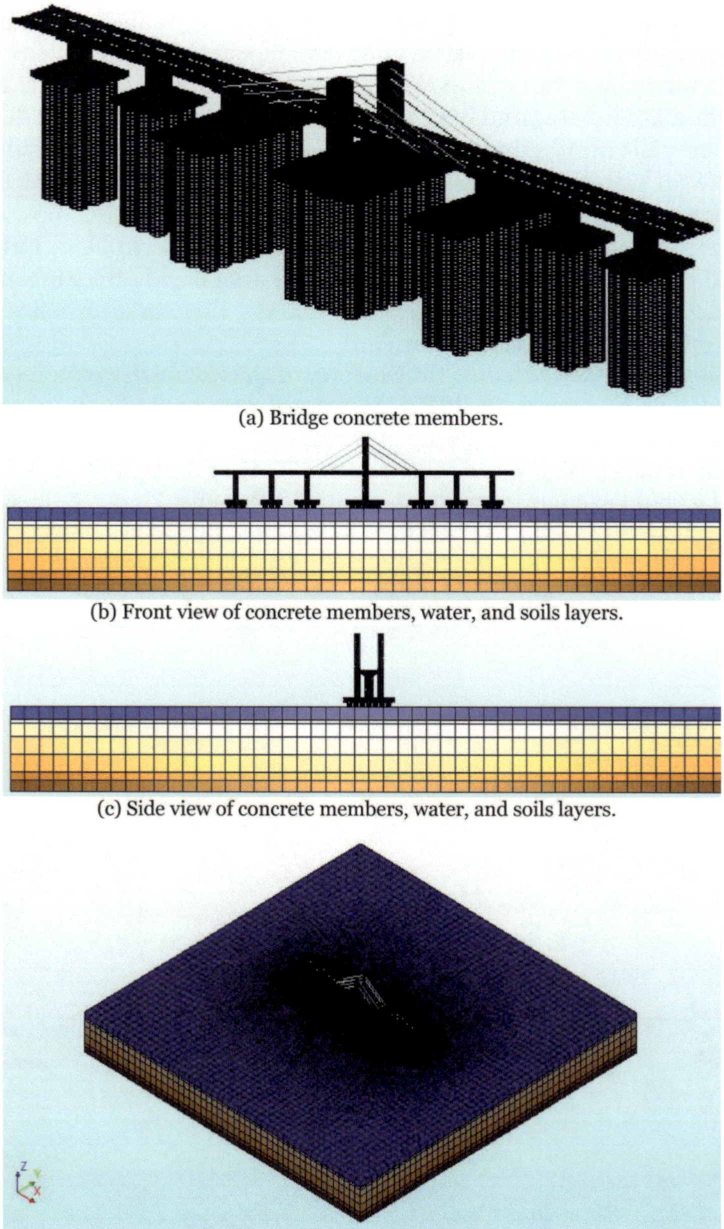

(a) Bridge concrete members.

(b) Front view of concrete members, water, and soils layers.

(c) Side view of concrete members, water, and soils layers.

(d) 3D view of the case study.

Figure 2.
Finite element mesh of the model.

the bedrock surface was selected based on a shear wave velocity exceeding 500 m/s. The model includes 2242 piles, distributed across beam elements, and comprises a total of 2280 nodes, capturing the dynamic properties under cyclic loads.

Figure 2 presents the finite element mesh of the entire system, showcasing the detailed configuration and distribution of the soil, piles, and pier components.

5. Test conditions

The exact dynamic response of the bridge pier to the combined effects of earthquake-induced ground motion, waves, and water currents was simulated. The 100-year current-wave characteristics of the field area are detailed in **Table 2**. **Figure 3** depicts the simulation setup and boundary conditions applied to the model. In the simulation, only the transverse (y-axis) direction of the bridge was considered for current and wave water loads. All simulations began with a water level of 12 meters.

By simulating these interactions, the model provides insights into how the bridge pier behaves under complex dynamic conditions, which is essential for understanding its structural resilience and for informing effective design and safety measures. It is possible to determine that $C_D = 1.2$, $C_M = 2.\ 0$, and $C_A = 1.0$. in accordance with the Environment Conditions and Load Standard.

In this chapter, the acceleration-time history of the Halabjah earthquake, which struck Baghdad on November 12, 2017 [35], was selected as the input seismic excitation to evaluate the structural response of the model. The earthquake's original accelerogram, with a peak ground acceleration (PGA) of 0.11 g at 41.5 seconds, spans a total ground excitation time history of 200 seconds. **Figure 4** illustrates the

Return period (Year)	Current velocity (m/sec)	Wave height (m)	Wave period (sec)	Wave length (m)	Wave velocity (m/sec)
100	2.5	0.5	4.5	8.5	4.5

Table 2.
Current-wave parameters.

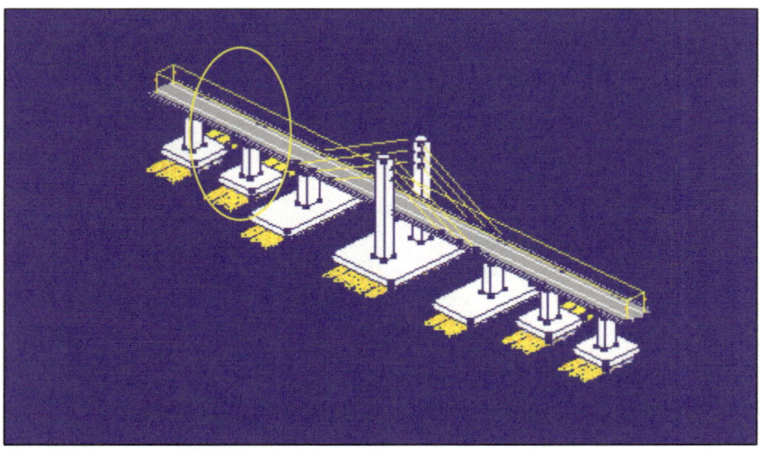

Figure 3.
Distributed impact loads along the build model.

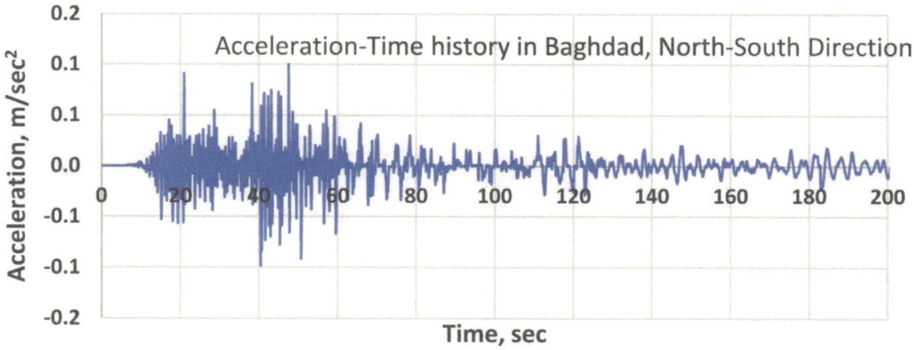

(a) Acceleration-Time history in Baghdad, North-South Direction.

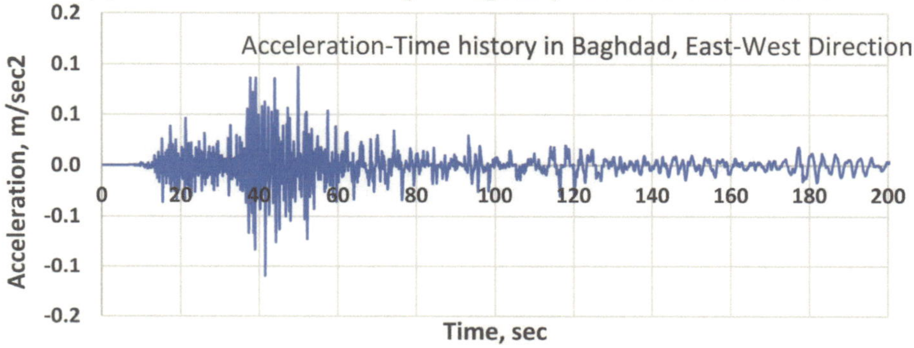

(b) Acceleration-Time history in Baghdad, East-West Direction.

Figure 4.
Acceleration-time history of Halabjah earthquakes in Baghdad.

acceleration-time histories for the longitudinal and transverse horizontal components of this seismic event. To investigate the impact of different earthquake motion amplitudes, three peak acceleration levels (0.05 g, 0.10 g, and 0.20 g) at a frequency of 20 Hz were considered. The finite element model was analyzed under deepwater conditions for the combined loads of current-wave and earthquake, taking into account the imposed boundary conditions.

6. Calculation results and analysis

6.1 Structural stability analysis

The analysis of the structural stability of the bridge pier begins with evaluating the displacement relative to the bottom of the pier and the absolute acceleration of the pier top. These metrics are crucial for understanding the pier's seismic response and overall structural stability.

The displacement of the pier top relative to its bottom is a critical factor in assessing earthquake-induced deformation in the pier's seismic design. Similarly, the absolute acceleration of the pier top significantly influences deck motion under bedrock ground motion. **Figure 5** presents the relative displacement of the pier and the peak acceleration resulting from earthquake ground movements at different peak ground accelerations (0.05 g, 0.10 g, and 0.20 g), with and without the influence of current and wave.

Hydrodynamic pressure significantly alters the pier's reactivity, reaching its peak near the top of the pier and affecting the relative peak displacement response. The data indicate that the peak relative displacement of the pier top increases gradually as the hydrodynamic pressure changes the pier's reactivity. The relative peak displacement under the combined current-wave and earthquake actions is greater than under earthquake alone. This increase in displacement response decreases with higher input accelerations from 0.05 g to 0.20 g. Specifically, with an input acceleration of 0.05 g, the peak relative displacement is more pronounced due to the current-wave impact,

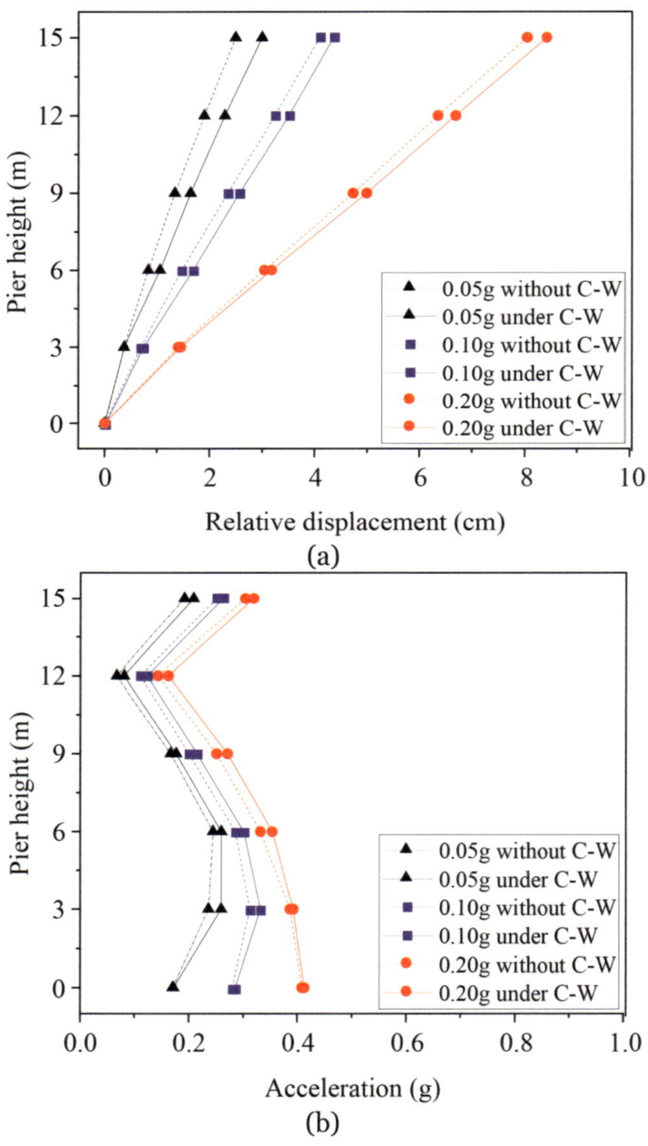

Figure 5.
Relative displacement and acceleration responses of the pier.

while for input accelerations of 0.10 g and 0.20 g, the effect of the current wave diminishes, as shown in **Figure 5(a)**.

Moreover, the bottom of the pier is subjected to greater earthquake forces, leading to concrete cracking and the formation of a plastic hinge. This reduces the flexural rigidity of the pier's bottom section, causing a decrease in acceleration and an increase in relative displacement. The pier's acceleration increases in the upper area due to the higher inertial impact of the structure. However, the acceleration at the top and bottom of the pier remains relatively homogeneous, as depicted in **Figure 5(b)**. In conclusion, the analysis shows that the acceleration at the top of the pier induced by the combined action of current waves and earthquakes is minimal compared to that under earthquake action alone.

6.2 Internal forces and plastic hinge formation

During an earthquake, the pier can sustain damage and develop a plastic hinge when the internal force response at a specific location exceeds its bearing capacity. In a single-column pier, the potential plastic hinge zone is typically located at the lowest portion of the pier. This chapter compares the shear force and moment at the pier bottom under earthquake action alone and under the combined actions of earthquake and current-wave impacts to analyze the effect on the pier's internal forces.

Figure 6 depicts the maximum shear force and moment response of the pier. It clearly illustrates that the shear force at the pier bottom is significantly greater than in the higher sections (**Figure 6a**). The substantial force acting on the pier bottom increases the likelihood of plastic hinge formation in this area during an earthquake. Additionally, the moment at the pier bottom within the first 6 meters is notably higher than at other locations along the pier (**Figure 6b**).

When subjected to a 0.05 g acceleration, the current-wave impact on the pile body is negligible. However, when the earthquake excitation increases to 0.10 g and 0.20 g, the forces acting on the pier body become more pronounced, indicating a visible increase in internal force responses under these conditions [36, 37]. This highlights the significant role that combined earthquake and hydrodynamic pressures play in the structural behavior and potential damage mechanisms of deepwater bridge piers.

During an earthquake, the plastic hinge area is readily produced at the bottom of the piers, where hydrodynamic pressure can have a significant impact. Therefore, for the seismic design of piers in deep water, it is crucial to account for the current-wave effects [38]. Several measures can be implemented to enhance the seismic performance of the piers under these conditions: Firstly, improved Connection of Longitudinal Steel: Extending the pier longitudinal steel to better connect with the cap or capping beam ensures that the strength of the steel is fully utilized, thereby improving the structural integrity and resilience of the pier under seismic loads. Secondly, use of Confining Stirrups: Implementing round or spiral stirrups with small spacing can effectively constrain the horizontal deformation of the core concrete. This confinement helps to maintain the structural integrity of the concrete under the combined actions of seismic and hydrodynamic forces.

By integrating these measures, the effects of hydrodynamic pressure can be more effectively addressed in the seismic design of deepwater bridge piers, thereby enhancing their stability and performance during earthquakes.

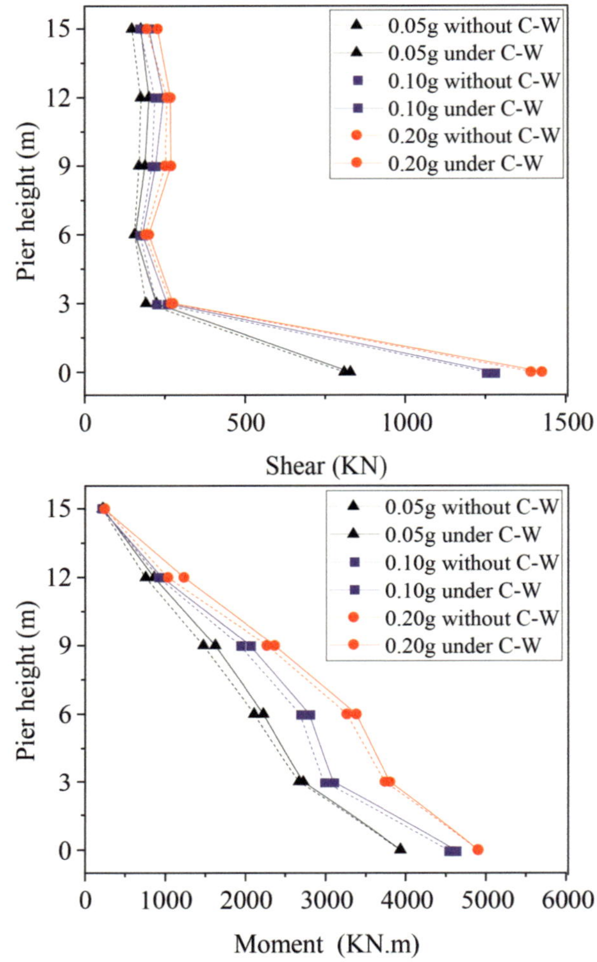

Figure 6.
Shear force and moment responses of the pier.

6.3 Earthquake hydrodynamic pressure influence coefficient

The hydrodynamic pressure influence coefficient, K, of the pier top's peak dynamic response under current-wave is definite as follows:

$$K = \begin{pmatrix} \textit{Dynamic response peak value under current, wave, earthquake} - \\ \textit{Dynamic response peak value under earthquake only} \end{pmatrix} / \quad (7)$$
$$(\textit{Dynamic response peak value under earthquake only}) \times 100\%$$

The K_d, K_a, K_s, and K_m shows the hydrodynamic pressure influence coefficient of the relative displacement, acceleration on the top of the pier, and the shear force, moment at the bottom of the pier separately. Therefore, the influence of hydrodynamic pressure on the peak dynamic response of bridge piers tends to emphasize peak acceleration and the interaction with earthquake waves, even though this influence can be less straightforward to discern. Across various parameters, the average impact reveals distinct trends:

Relative displacement experiences minimal alteration due to hydrodynamic pressure, indicating that its influence is negligible in comparison to other factors. In contrast, acceleration at the top of the pier shows a significant response, with hydrodynamic effects contributing up to a maximum of 11%. This underscores the substantial impact of currents and waves on the pier's acceleration dynamics during seismic events.

Meanwhile, the influence on shear force and moment at the bottom of the pier is more modest, averaging around 4%. Although these forces are affected by hydrodynamic pressures, the magnitude of their alteration remains relatively consistent across different scenarios. This consistency suggests that while hydrodynamic factors do affect shear and moment, their overall effect is less pronounced compared to acceleration.

These findings underscore the critical role of earthquake ground motion characteristics in determining how currents and waves influence the dynamic behavior of bridge piers. By understanding these interactions, engineers can better design and reinforce piers to withstand the combined challenges of seismic activity and hydrodynamic forces, ensuring robust structural performance in deepwater environments.

7. Coastal bridge design: practical implications

The findings from the structural stability analysis of bridge piers under combined seismic and hydrodynamic forces reveal important insights for enhancing engineering practices. By delving into the practical applications of these findings, we can significantly influence engineering practices, safety standards, and bridge design principles in seismically active coastal regions.

Engineers should incorporate hydrodynamic pressure considerations into seismic design models. This involves using computational fluid dynamics (CFD) simulations to estimate hydrodynamic forces and integrating these forces into structural analysis software. The application of improved longitudinal steel connections and the use of confining stirrups can significantly enhance the strength of bridge piers. This will require revising standard reinforcement detailing practices to include these techniques as mandatory for coastal bridges.

Engineers should regularly evaluate the dynamic responses of bridge piers to varying peak ground accelerations. This can be achieved by implementing advanced structural health monitoring systems that continuously measure and analyze pier responses during seismic events and storms. By understanding these dynamic responses, engineers can adjust design parameters such as damping ratios and natural frequencies to improve the performance of bridge piers under combined loading conditions.

Design codes should be updated to include guidelines for evaluating the combined effects of seismic and hydrodynamic forces. This means specifying design loads that reflect the interaction of these forces, ensuring that safety margins are adequate for both seismic activity and wave impacts. Transitioning from prescriptive to performance-based design codes can provide more flexibility in achieving safety objectives, allowing for innovative solutions tailored to specific site conditions.

Safety standards should mandate regular inspections that specifically look for damage caused by seismic and hydrodynamic forces. This can involve using non-destructive testing methods such as ground-penetrating radar and ultrasonic testing. Maintenance protocols should address identified vulnerabilities, including corrosion of reinforcement due to seawater exposure and structural damage from seismic activity. This ensures the long-term durability and functionality of bridge piers.

Engineers should adopt a performance-based design approach that sets specific objectives for pier performance under combined loading conditions. This includes defining acceptable levels of damage and functionality post-event. Exploring solutions such as seismic isolation systems, which decouple the pier from ground motion, and energy-dissipating devices, which absorb and dissipate energy from both seismic and hydrodynamic forces, can enhance resilience.

Future research should focus on developing advanced simulation techniques that accurately model the interaction between seismic and hydrodynamic forces. This includes using multi-physics software that can simulate complex interactions. Conducting field studies to validate simulation results and refine design criteria is essential. This can involve monitoring the performance of existing coastal bridges during seismic and storm events to gather real-world data.

By expanding on these practical implications, we can ensure that the research not only advances theoretical knowledge but also provides tangible benefits for practitioners and policymakers, ultimately leading to safer and more resilient coastal bridge structures.

8. Expanding the scope of case studies or simulations

Incorporating a broader range of case studies or additional simulations is essential for gaining a more nuanced and comprehensive understanding of how seismic and hydrodynamic forces impact bridge structures in different settings. By extending the scope beyond a single case study or simulation, researchers can explore how various geographical and environmental factors influence the behavior of bridge piers under combined seismic and hydrodynamic loads. This expanded approach not only enriches the research findings but also improves the generalizability of the results, making them applicable to a wider range of real-world scenarios.

8.1 Understanding variations across different conditions

A single case study or simulation often reflects a specific set of conditions, such as the seismic activity levels, hydrodynamic forces, and environmental settings present at that location. By broadening the range of case studies, researchers can examine how these factors vary across different regions and conditions. For instance, bridges in coastal regions might face different wave patterns, tidal ranges, and seismic risks compared to those in other locations. By including a variety of case studies, researchers can identify patterns and differences in bridge responses, such as how varying intensities of waves and different seismic magnitudes affect the stability and performance of bridge structures. This approach helps to uncover potential vulnerabilities and design challenges that may not be evident from a single case study.

8.2 Enhancing the generalizability of findings

Expanding the scope of research through diverse case studies and simulations enhances the generalizability of the findings. While a specific study provides valuable insights into the conditions of a particular site, a broader approach allows for the examination of how well the results apply to different environments and scenarios. For example, by analyzing bridge piers under various earthquake magnitudes, hydrodynamic pressures, and geographical settings, researchers can develop more universally applicable design principles and safety guidelines. This broader perspective helps to ensure that the design recommendations and engineering practices developed from the study are robust and effective across a range of conditions, rather than being limited to the specifics of one case.

8.3 Application to design and safety standards

The insights gained from a more comprehensive set of case studies can directly inform the development of improved design standards and safety protocols for bridge construction and maintenance. For example, if the expanded research reveals that certain design features perform well across different conditions, these features can be incorporated into standard design practices. Conversely, if the research uncovers new challenges or limitations, these findings can lead to revisions in safety standards and design guidelines. Ultimately, a broader scope of research helps engineers and policymakers develop more effective and adaptable strategies for designing resilient bridge structures capable of withstanding diverse environmental and seismic conditions.

9. Practical implications for bridge design and maintenance in seismic zones

9.1 Design modifications for seismic resilience

The research findings emphasize the importance of incorporating specific design modifications to enhance the seismic resilience of bridges in seismically active zones. These modifications aim to address the combined effects of seismic forces and hydrodynamic pressures on bridge piers and other structural elements. Here are several design modifications based on the study's results:

Extend the longitudinal steel reinforcement within bridge piers to improve the connection between the pier and the cap beam. By increasing the length and strength of the steel reinforcement, engineers can better transfer seismic forces from the pier to the cap beam, reducing the risk of structural failure. This modification helps to maintain the structural integrity of the bridge during strong seismic events and prevents premature damage.

Implement round or spiral stirrups with smaller spacing in the core concrete of bridge piers. These stirrups provide additional confinement to the concrete, enhancing its ductility and strength under seismic loads. This design feature helps to manage deformation and reduce the likelihood of cracks forming in the concrete, which can weaken the structure during an earthquake.

Incorporate seismic isolation devices such as elastomeric bearings or sliding bearings into bridge designs. These systems absorb and dissipate seismic energy, reducing the forces transmitted to the bridge structure. Seismic isolation systems can help to limit the amount of movement and damage experienced by the bridge during an earthquake, improving its overall resilience.

Design piers with features that address the effects of hydrodynamic pressures from currents and waves. For example, engineers can use tapered or streamlined pier shapes to reduce drag and minimize the impact of wave forces. Additionally, designing for the worst-case hydrodynamic scenarios ensures that piers can withstand both seismic and wave-induced loads.

These design modifications are directly informed by the study's findings and provide practical solutions for engineers to enhance the seismic performance of bridges in seismic zones.

9.2 Maintenance strategies for long-term durability

In addition to design modifications, effective maintenance strategies are crucial for ensuring the long-term durability and safety of bridges in seismically active areas. The following maintenance practices are recommended based on the study's insights:

Implement a routine inspection schedule to assess the condition of bridge piers and other critical components. Inspections should focus on detecting signs of damage such as cracks, deformations, and corrosion. Regular assessments help identify issues early and prevent minor problems from escalating into significant structural failures.

Install monitoring systems to measure hydrodynamic pressures and currents affecting the bridge piers. These systems can provide real-time data on environmental conditions and help engineers evaluate how well the bridge is performing under current and wave forces. Data from these systems can guide maintenance decisions and inform future design improvements.

For bridges equipped with seismic isolation systems, regular maintenance of these devices is essential. Engineers should inspect isolation bearings for wear and tear, ensure that damping mechanisms are functioning correctly, and perform necessary repairs or replacements. Maintaining these systems ensures they continue to provide effective protection against seismic forces.

As new research findings become available, maintenance protocols should be updated to incorporate the latest knowledge and best practices. Staying informed about advancements in bridge engineering and applying new insights can lead to improved maintenance strategies and enhanced bridge performance.

These maintenance strategies help to ensure that bridges remain safe and functional over time, addressing both existing and emerging challenges in seismically active regions.

9.3 Development of comprehensive design guidelines and standards

The research findings also highlight the need for the development of comprehensive design guidelines and safety standards for bridge construction in seismic zones. These guidelines should include:

- *Detailed Design Criteria:* Develop detailed design criteria that incorporate the combined effects of seismic forces and hydrodynamic pressures. Guidelines should specify design requirements for structural elements, including the use of specific materials, construction techniques, and safety factors.

- *Best Practices for Seismic and Hydrodynamic Design:* Establish best practices for engineers to follow when designing bridges for seismic and hydrodynamic forces. These practices should be based on the latest research findings and include recommendations for addressing common design challenges.

- *Guidance for Risk Assessment and Management:* Provide guidance on how to assess and manage risks associated with seismic and hydrodynamic forces. This includes methods for evaluating potential risks, developing risk mitigation strategies, and preparing for emergency response in the event of a seismic event.

By developing these guidelines and standards, engineering practices can be standardized to ensure that all new bridge projects meet high safety and performance criteria.

10. Conclusion

This study thoroughly investigated the dynamic response of pile foundation bridge piers under the combined influence of earthquake ground motions and hydrodynamic pressures from currents and waves. Utilizing the Halabjah earthquake excitation, Morison hydrodynamic pressure formula, and nonlinear models of soil and concrete, the research comprehensively assessed the structural behavior of bridge piers subjected to these complex loading conditions. The following conclusions were drawn:

1. The study revealed that under combined current-wave and earthquake actions, the pier's relative peak displacement was significantly higher compared to conditions with only earthquake forces. This underscores the substantial influence of hydrodynamic pressures on the displacement dynamics of bridge piers during seismic events.

2. The analysis indicated that while hydrodynamic effects had a noticeable impact on acceleration, particularly at the pier top, the magnitude of this influence was less pronounced compared to relative displacement. Hydrodynamic pressures contributed up to an 11% variation in acceleration, highlighting their role in modifying the dynamic response but indicating a more moderate influence compared to displacement.

3. Hydrodynamic pressures also affected the shear force and moment at the bottom of the pier, with an average influence of around 4%. This suggests that while these forces are influenced by currents and waves, their overall alteration remains relatively consistent across different scenarios.

4. The study established a clear connection between the characteristics of earthquake ground motions and the effects of current-wave loads on the pier's dynamic response. This relationship underscores the importance of considering seismic and hydrodynamic interactions in the design and analysis of pile foundation bridge piers.

5. It is imperative to incorporate current-wave effects into the seismic design criteria for pile foundation bridge piers spanning seas and oceans. Neglecting these hydrodynamic forces can compromise the safety and reliability of such structures during seismic events.

Further research is needed to enhance the understanding of hydrodynamic pressure effects on bridge dynamics. Future studies should focus on refining models to better simulate the complex interactions between structural systems and random actions such as currents, waves, and earthquakes. This will improve the accuracy and reliability of dynamic response predictions and seismic analyses for bridge structures.

Acknowledgements

The corresponding author extends his thanks to Al- Mustaqbal University for the financial support for the research.

Author details

Riyadh Alsultani[1*], Ibtisam R. Karim[2], Saleh I. Khassaf[3] and Ahmed Ashor Al-Saadi[4]

1 Civil Engineering Department, College of Engineering, University of Babylon, Babylon, Iraq

2 Civil Engineering Department, University of Technology, Baghdad, Iraq

3 Civil Engineering Department, University of Basrah, Basrah, Iraq

4 Department of Civil Engineering, College of Engineering, Al-Qasim Green University, Babylon, Iraq

*Address all correspondence to: dr.riyadh.abdulabbas@uomus.edu.iq

References

[1] Huang S, Liu C. Dynamic behavior analysis of bridge pier under impact of dam-break flood in different directions. Natural Hazards. 2024;**120**(3):2705-2730

[2] Alsultani R, Karim IR, Khassaf SI. Dynamic response of Deepwater pile foundation bridge Piers under current-wave and earthquake excitation. Engineering and Technology Journal. 2022;**40**(11):1589-1604. DOI: 10.30684/etj.2022.135776.1285

[3] Alsultani R, Karim IR, Khassaf SI. Dynamic response analysis of coastal piled bridge pier subjected to current, wave and earthquake actions with different structure orientations. International Journal of Concrete Structures and Materials. 2023;**17**(1):1-5. DOI: 10.1186/s40069-022-00561-5

[4] Alsultani R, Karim IR, Khassaf SI. Mathematical formulation using experimental study of hydrodynamic forces acting on substructures of coastal pile foundation bridges during earthquakes: As a model of human bridge protective. Resmilitaris. 2022;**12**(2):6802-6821

[5] Li ZX, Zheng Q, Su J, Shi Y, Zhao B. Underwater shaking table tests of a sea-crossing cable-stayed bridge under combined earthquake and wave action. Ocean Engineering. 2023;**287**:115871

[6] Yamada Y, Kawano K, Iemura H, Venkataramana K. Wave and earthquake response of offshore structures with soil-structure interaction. Doboku Gakkai Ronbunshu. 20 Oct 1988;**1988**(398):157-166

[7] Karadeniz H. Spectral analysis of offshore structures under combined wave and earthquake loadings. In: 9th International Conf. On Offshore and Polar Eng. Brest: ISOPE; 1999. pp. 504-511

[8] Fukusumi T, Uchida N, Takahashi D. Response characteristics of soft settled type of offshore structures subjected to Harmonic Sea wave and earthquake wave. In: 13th International Conf. On Offshore and Polar Eng. Honolulu: International Society of Offshore and Polar Engineers; 2003. pp. 1007-1012

[9] Etemad AK, Gharabaghi ARM, Chenaghlou MR. Nonlinear dynamic behavior of fixed jacket-type offshore platforms subjected to simultaneously acting wave and earthquake loads. In: 23rd International Conf. On Offshore Mech. And Arctic Eng. Vancouver: American Society of Mechanical Engineers; 2004. pp. 893-900

[10] Abbasi M, Gharabaghi ARM. Study the effect of wave directionality on dynamic nonlinear behavior of jack-up subjected to wave and earthquake loading. In: 26th International Conf. On Offshore Mech. And Arctic Eng. San Diego: American Society of Mechanical Engineers; 2007. pp. 377-384

[11] Furong L, Guoxing C, Zhihua W. Seismic responses of single-column pier considering the effects of hydrodynamic pressure. Journal of Earthquake Engineering and Engineering Vibration. 2008;**28**(2):55-59

[12] Xiaoyu H, Hongnan L. Research on piles with small dimension under the combination of seismic and wave action. Journal of Earthquake Engineering and Engineering Vibration. 2007;**27**(5):139-145

[13] Song B, Zheng F, Li Y. Study on a simplified calculation method for

hydrodynamic pressure to slender structures under earthquakes. Journal of Earthquake Engineering. 2013;**17**(5):720-735

[14] Al-Taie AJ, Albusoda BS. Earthquake hazard on Iraqi soil: Halabjah earthquake as a case study. Geodesy and Geodynamics. 2019;**10**(3):196-204

[15] Wei C, Zhou D, Ou J. Wave and wave-current actions on a bridge tower: An experimental study. Advances in Structural Engineering. 2019;**22**(6):1467-1478

[16] Huynh LE, Chu CR, Wu TR. Hydrodynamic loads of the bridge decks in wave-current combined flows. Ocean Engineering. 2023;**270**:113520

[17] Chen Y, Huang X, Wu K, Li ZX. Experimental research on dynamic responses and hydrodynamic pressures of deep-water bridge piers under seismic and wave actions. Engineering Structures. 2024;**313**:118276

[18] Zhou S, Zhai G. A multi-Hazard risk assessment framework for urban disaster prevention planning: A case study of Xiamen, China. Land. 2023;**12**(10):1884

[19] Kenarkoohi M, Hassan M. Review of accelerated construction of bridge piers-methods and performance. Advances in Bridge Engineering. 2024;**5**(1):3

[20] Malerba PG. Bridge vulnerabilities and collapses: The Italian experience. Structure and Infrastructure Engineering. 2024;**20**(7-8):976-1001

[21] Xu G, Cai CS. Numerical simulations of lateral restraining stiffness effect on bridge deck–wave interaction under solitary waves. Engineering Structures. 2015;**101**:337-351

[22] Qu K, Yao W, Tang HS, Agrawal A, Shields G, Chien SI, et al. Extreme storm surges and waves and vulnerability of coastal bridges in New York City metropolitan region: An assessment based on hurricane Sandy. Natural Hazards. 2021;**105**:2697-2734

[23] Wei K, Yuan W, Bouaanani N. Experimental and numerical assessment of the three-dimensional modal dynamic response of bridge pile foundations submerged in water. Journal of Bridge Engineering. 2013;**18**(10):1032-1041

[24] Ministry of Communications of China. Code for Design of Ground Base and Foundation of Highway Bridges and Culverts. Beijing: JTG D63-2007, China Communications Press; 2007 (in Chinese)

[25] Chen GX, Zhuang HY. Developed nonlinear dynamic constitutive relations of soils based on Davidenkov skeleton curve. Journal of Geotechnical Engineering. 2005;**27**:860-864, in Chinese

[26] Lee J, Fenves GL. Plastic-damage model for cyclic loading of concrete structures. Journal of Engineering Mechanics. 1998;**4**:892-900

[27] Huang X, Chen L, Gao Y. Nonlinear random vibration of the slender deep-water pier under seismic excitation. Probabilistic Engineering Mechanics. 1 Apr 2023;**72**:103423

[28] Morison JR, O'Brien MP, Johnson JW, Schaaf SA. The force exerted to surface wave on piles. Petroleum Transaction, AME. 1950;**189**:149-154

[29] Teng CC, Nath JH. Forces on horizontal cylinder towed in waves. Journal of Waterway, Port, Coastal, and Ocean Engineering. 1985;**111**(6):1022-1040

[30] Alfatlawi TJ, Alsultani RA. Characterization of chloride penetration in hydraulic concrete structures exposed to different heads of seawater: Using hydraulic pressure tank. Engineering Science and Technology, an International Journal. 2019;**22**(3):939-946. DOI: 10.1016/j.jestch.2019.02.001

[31] Thair JM, Imad AD, Riyadh AA. Experimental determination and numerical validation of the chloride penetration in cracked hydraulic concrete structures exposed to severe marine environment. IOP Conference Series: Materials Science and Engineering (Istanbul, Turkey). 2018;**454**:012099. DOI: 10.1088/1757-899X/454/1/012099

[32] Alfatiawi T, Mansori N, Alsultani A. Stability assessment of Diaphram cellular cofferdams subjected to severe hydro-structural conditions. Open Civil Engineering Journal. 2020;**14**(1):44-55

[33] Alfatlawi TJ, Alsultani RA. Numerical modeling for long term behavior of chloride penetration in hydraulic concrete structures. Global Scientific Journal of Civil Engineering. 2018;**1**:6-10

[34] Ali Al-Sultani RAA, Salahaldain Z, Naimi S. Features of monthly precipitation data over Iraq obtained by TRMM satellite for sustainability purposes. Al-Mustaqbal Journal of Sustainability in Engineering Sciences. 2023;**1**(2):2

[35] Alsultani R, Khassaf SI. Nonlinear dynamic response analysis of coastal pile foundation bridge pier subjected to current, wave and earthquake actions: As a model of civilian live. Resmilitaris. 2022;**12**(2):6133-6148

[36] Alfatlawi TJM, Alsultani RAA. Determination of the degree of saturation and chloride penetration in cracked hydraulic concrete structures: Using developed electrical conductivity technique. Indian Journal of Science and Technology. 2018;**11**:37

[37] Salahaldain Z, Naimi S, Alsultani R. Estimation and analysis of building costs using artificial intelligence support vector machine. Mathematical Modelling of Engineering Problems. 2023;**10**(2):405-411. DOI: 10.18280/mmep.100203

[38] Mazinani I. Experimental and Computational Studies of Fluid Structure Interaction of Tsunami Bore on Coastal Bridges [Doctoral dissertation]. Malaysia: University of Malaya; 2021

Chapter 4

Effects of Backfill Types on the Behavior of Buried Concrete Sewerage Pipelines in Soil

Nariman Hisham Halalo

Abstract

Pipelines are commonly buried underground to provide environmental stability, temperature insulation and mechanical protection. Pipelines serve as effective underground infrastructure especially as sewerage pipeline. Today, it is well known that besides the pipe material, the installation procedures have a great effect on the performance of the pipeline in soil. This paper explores the performance of sewerage pipeline installed in different soil according to Syria procedures with compacted and uncompacted backfill materials (ML, CH) and subjected to different loadings to attempt for understanding the effects of backfill types on the behavior of rigid sewerage pipelines using three dimensional finite element approaches (ABAQUS). The effects of different parameters are studied such as, depth of backfill, and backfill compaction. In this model, the material behavior of compacted or uncompacted soil is described using an elastic-plastic constitutive model with Mohr-Coulomb failure criterion. The objectives of the current study are to study the effects of the backfill compaction conditions on the behavior of a buried concrete sewerage pipeline in soil through numerical model.

Keywords: buried pipelines, infrastructure, backfill compaction, Finite element method (ABAQUS), Mohr-Coulomb material model

1. Introduction

Pipelines are common transportation means for sewerage water, drinking water, oil and natural gas. Generally, these pipelines laid underground for economic, esthetic, safety, and environmental reasons. It is indisputable that sewerage networks are an essential component of sustainable development, and greatly affect the health and safety of people all over the world.

The degradation of pipelines over time due to the effects of soil overlay along with the effects of backfill material conditions can sometimes cause catastrophic failures in sewerage pipelines. And the need to use computer programs to solve geotechnical and construction issues increases with the increase in urban development and growth. Present study has focused on the failure criteria of the buried pipeline under load conditions. A model in the platform of ABAQUS has been developed for this purpose. Investigation has been carried out for different soil types and finally the model has been validated with a reference work [1].

Underground pipeline failure occurs when applied stresses exceed its structural elasticity. Technical analysis for proper design of buried sewerage pipelines should take into account the characteristics of these pipelines, internal and external loads, and surrounding conditions such as backfill materials, installation depth, compaction quality, and road superstructure loads [2].

This paper explores the performance of sewerage pipeline installed in different Backfill soil conditions according to Syria procedures with compacted and uncompacted backfill materials (ML, CH) [3] and subjected to different loadings to study the effects of the backfill compaction with different burial depth (1, 2, 3, and 4 m) on the buried sewerage pipeline. Finite element analysis (FEA) will discuss in detail the depths of bury depth (1, 2, 3, and 4 m). The main objectives of the research study are to study the effects of the backfill compaction conditions on the behavior of a buried concrete sewerage pipeline in soil through numerical model. Several major points in the soil and pipeline will be taken to record the displacements and stresses in the soil and pipeline.

2. Background

Buried pipelines can be classified as either flexible or rigid depending on whether the pipeline itself can deform up to 2% without incurring damage. For rigid pipelines, the bottom of the pipeline should be supported. As for flexible pipelines, both bottom and side parts of the pipeline must be supported. Rigid pipelines, such as reinforced/non-reinforced concrete, and clay pipelines are subject to significant structural cracking if deflection is greater than 2% [4]. Proper pipeline installation involves much more than just covering up the pipeline. A buried pipeline is a structure that incorporates both the properties of the pipelines and the properties of the soil surrounding the pipelines. The geotechnical-structural design of a pipeline is based on certain soil conditions, and construction control is important to ensure these conditions are applied.

2.1 Soil terminology

Bedding is the area at the bottom of the pipeline inside the trench.

Embedment is the soil placed around the pipeline.

Backfill is the material used to refill the trench above the pipeline [5], sometimes the term controlled low strength material (CLSM) is also used to describe flowable backfill (**Figure 1**) [6].

2.2 Pipeline terminology

The pipeline is divided into five parts, these parts are: crown, invert, springline, pleura, haunch. The crown which is the top of the pipeline, the invert is the bottom of the pipeline, the springline which is in the middle of the pipeline, the supporting upper part above the middle line of the pipeline is called pleura. The lower supporting part is called the haunch (**Figure 2**) [5].

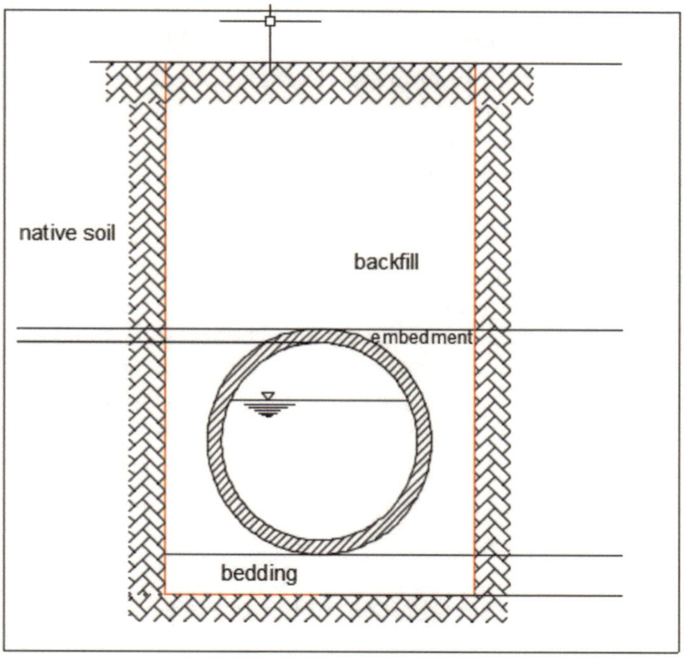

Figure 1.
Soil trench terminology.

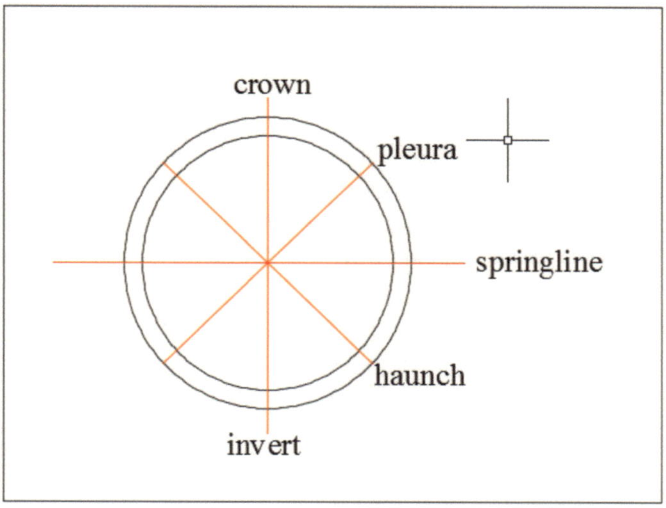

Figure 2.
Pipeline terminology.

3. Numerical studies in the literature

In fact, there are many reference studies that deal with the subject of the studied research, and among these studies, the researcher Terzi et al. [7] buried HDPE pipelines, the soil was poorly graded sand with different compaction soil degrees to measure the strain response of pipeline. The study's results indicated that when the backfill soil which is placed around the pipeline is loose, the maximum strain occurs at the invert of the pipelines, also when the backfill soil is fully compacted, the maximum strain occurs on the crown of the pipelines [7]. Another researcher Brachman et al. [8] conducted laboratory experiments to study the effects of compaction soil degrees, backfill materials, and gradation on the deflection, strain, and local bending of special-shaped thermoplastic pipelines. They proposed an empirical coefficient based on measured data to simplify the maximum bending strain calculation in designing these pipelines [8].

The preceding review shows that finite element methods effectively can be used to study and analyze the performance of buried pipelines. These studies have provided a database for the behavior and performance of buried pipelines in certain field conditions in Syria. Further studies may be needed in order to better understand the overall field performance of buried pipelines, and the effects due to changes in Backfill soil conditions. The geometric configuration of the pipeline trench, including the cover height and trench width, significantly influenced the overall deformations of the pipeline system.

4. Statement of the problem

The environmental conditions were not fully considered in the pipeline analysis procedure. Thus embedment zone that surrounding the pipeline with 30 cm granular material (Class II soil of ASTM D2321) (**Figure 3**) [9].

The main reasons for the failure of the sewerage pipeline when performing groove backfilling, the bottom area of the pipeline is narrow, and there is a dead backfill angle in the effective support foot area of the pipeline. Because pressing equipment is difficult to use, it is difficult to guarantee the quality of pressing. A common knowledge nowadays in Syria that pipeline installation currently does not have soil consolidation tools for the installation of sewerage pipelines. Choosing an appropriate type of foundation layer depending on the material of the pipeline (rigid or flexible), choosing an appropriate type of backfill material. Depending on the soil conditions compacted or uncompacted soil, with suitable compaction at the site. Under the soil, so the effect of external loads such as soil weight and traffic loads becomes very important. The calculation will be done with backfilling of different types of pipes at different depths, under two conditions:

1. Compacted soil (to simulate the method of executing backfilling according to standards).

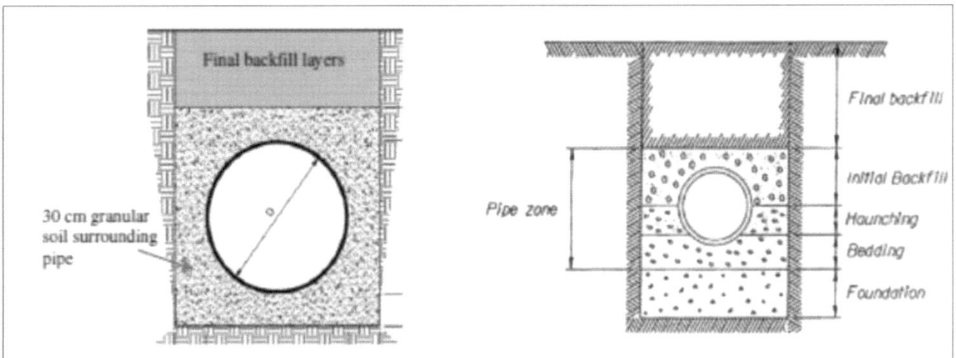

Figure 3.
Trench conditions according to Syria procedure typical pipeline backfilling according to ASTM2321.

2. Uncompacted soil (to simulate the backfill method used in projects).

These images show how sewerage pipelines are being implemented in some of the towns in Syria (**Figures 4** and 5).

Figure 4.
The buried sewerage concrete pipeline is being implemented in the town of Jabab in Dara-Syria.

Figure 5.
The buried sewerage HDPE pipeline is raising the sewerage outfalls of the Barada River in Syria.

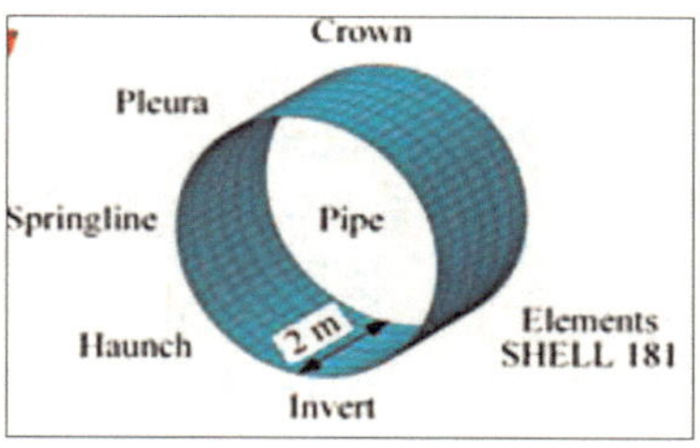

Figure 6.
The pipeline terminology used in the numerical model.

Several main points will be taken in the soil and the pipeline, and by collecting them a clear concept is obtained to study the effect of the specific parameters of the used backfill condition on the transitions (shifts) and stresses in the soil and the pipe. These points are: crown, pleura, springline, haunch, invert (**Figure 6**).

5. Units

The first task to do is to decide on the units to be used in the analysis. ABAQUS does not have any inherent set of units (**Table 1**) [10].

Length	Force	Stress	Time	Density
m	KN	Kpa	sec	KN/m^3

Table 1.
The units used in the numerical model.

6. Finite element modeling

3D finite element modeling was performed using ABAQUS software [10]. Shell elements were used for the buried and solidified elements of the surrounding soil.

6.1 Constitutive models

The dimensions of the model (the extension of the foundation soil) will be chosen largely to cover the area of stress distribution in the soil and the fading of reactions at the edges of the model, as the result of the reference research shows that the ideal dimensions for the numerical study of the case of a buried pipeline within a trench have a width of (B) so that the width of the trench is not less than (1.5d) according to AASHTO [11] and the width of both sides of the pipeline is not less than (0.5 m) according to (ASTM) [12–14] and therefore the width of the trench studied is (B = 1.3 m), and by adopting the soil of its horizontal extension (5B + B + 5B = 11B) and its depth (5B) from the bottom of the excavation level, given the depth of the excavation (H) as shown in the figure where B: width of the excavation (trench), H: depth of the excavation, d: diameter of the outer pipeline (**Figure 7**).

6.2 Types of elements used in the numerical model

Two types of elements are used in the present model. Soil has been used as 3-dimensional solid element with 20 nodes fully integration and pipeline has been modeled as shell element with 8 nodes fully integration. **Table 2** describes the element types, element shapes, geometric orders and number of elements in the model. Soil and concrete pipeline has been design in ABAQUS using two different parts. A 100 m long soil block has been designed as 3D deformable solid model with dimension 100 m × 15 m × 12 m. A 100 m long pipeline has been designed as a 3D deformable shell model with the diameter of (0.3 m) and thickness of (0.04 m) (**Figure 8**).

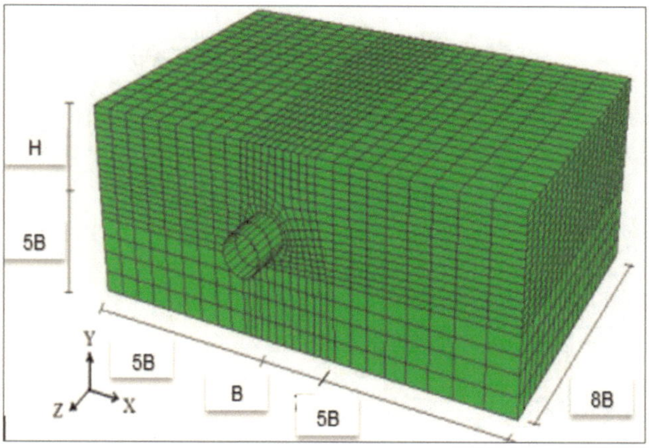

Figure 7.
The finite element model geometry in ABAQUS.

Name of element	Element types	Element shapes	Geometric orders	Geometric forms	Element numbers
Soil	C3D20R	hexahedral	quadratic		792
Pipeline	S8R	quadrilateral	quadratic		264

Table 2.
The types of elements used in the numerical model.

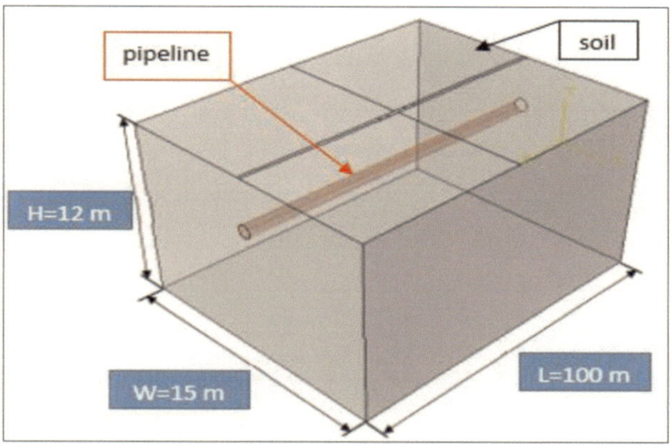

Figure 8.
Geometry of the numerical model designed in ABAQUS.

6.3 Material properties used for the simulations

The material properties of surrounding soils and pipeline used in the simulations were obtained from published literature.

6.3.1 Soil properties

There exist several types of soils: Semi hard Silty Clay Soil (CH) and Firm Sandy Silty Soil (ML) [3]. The current study was conducted for two types of soils and according to ASTM D2321 [9] pipeline regulations, the design criteria focus the installation of pipelines on the soil area surrounding the pipeline.

Table 3 describes the types of soils and their physical and mechanical properties used in the numerical model, and **Table 4** describes the properties of the pipeline used in the numerical model (**Figure 9**).

6.3.2 Pipeline properties

In order to investigate the effect of pipeline stiffness on soil behavior, the type of pipeline will be designed is concrete.

6.4 Soil behavior modeling

The most important factor in numerical modeling of the pipeline-soil interaction is simulating the stress-strain behavior of the soil [15]. Mohr-Coulomb model suitable for soil behavior model. The parameters must be defined required to determine the MC model, namely Young's modulus (E), Poisson's ratio (ν), angle of internal friction (Ø), volumetric expansion angle (ψ), soil volumetric weight (γ), and soil cohesion (c) [16].

Elastic properties	Type S1		Type S2		Unit
Soil site	Umm Riwaq—Suwayda—Syria		Deir Al-Hajar—Al-Ghazlaniya—Rural Damascus—Syria		—
Type of soil	Semi hard silty clay soil		Firm sandy silty soil		—
Soil classification	CH		ML		—
Unit weight (ϒ)	16.2		15.7		KN/m^3
Max density	14.2		15.5		KN/m^3
Plastic properties	Compacted soil	Uncompacted soil	Compacted soil	Uncompacted soil	—
Cohesion soil (C)	1	0.4	44	12	KN/m^2
Friction angle (ϕ)	25.9	22.8	33.9	12.8	(°)
Young's modulus	22769.6	7280.4	44,507	14,387	KN/m^2
Poisson's ratio (ν)	0.2	0.2	0.25	0.25	—
Dilation angle (ψ)	2	2	2	2	(°)

Table 3.
The types of soils and their physical and mechanical properties used in the numerical model.

Pipeline material type	Volumetric weight (KN/m^3)	Modulus of elasticity (Mpa)	Poisson's ratio	Resistance stress (Mpa)	Pipeline diameter (m)	Pipeline thickness (m)
Concrete pipeline	24	31,000	0.2	18	0.3	0.04

Table 4.
The properties of the pipeline used in the numerical model.

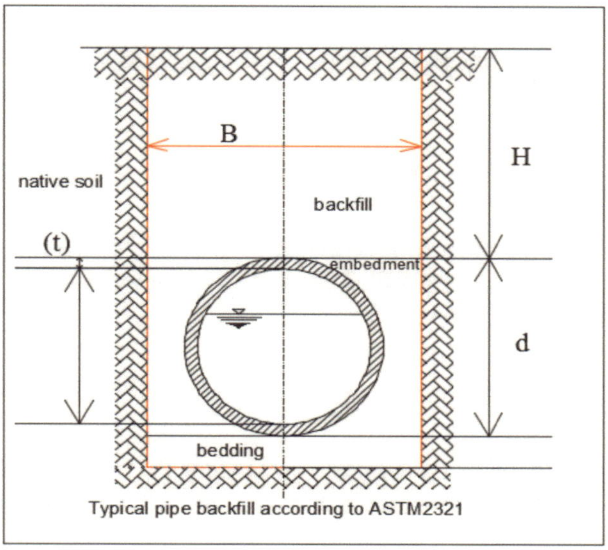

Figure 9.
Typical pipeline backfill according to ASTM2321 [9].

6.5 Interaction between soil and pipeline

Present study has been carried out for two types of soil. The soil and pipeline both have been studied as failure conditions, so selected material of pipeline is perfectly plastic and chosen material of soil is elastic-plastic material according to Mohr-Coulomb theory.

The interaction surfaces between the soil and the pipeline are created within the available surface contact method property (master and slave), There are several types of constraints to connect a solid body with a cortex. The elements are completely linked using the tie constraint, which connects two separate surfaces together so that there is no relative movement between them (all degrees of freedom are associated with each other at the point of convergence between the two elements linked together). The upper surface of the pipeline with the backfill above the pipeline, the lower surface of the pipeline with the pad under the pipeline, and the backfill surface and the upper surface of the pad with the lower surface of the backfill above the pipeline are joined.

6.6 Assembly and boundary conditions

The model is created by assembling the positioned components of the soil and pipelines, and the boundary condition of the end surfaces of the pipeline and the soil in all directions are created by assembling (x = y = z = 0).

6.7 Step and loading of the model

Table 5 describes the two types of loads have been applied to the model for the analysis. All the loads applied on the model have been already discussed section. The self-weight of the soil which is generally called dead load is represented as

Loads	Values	Units
Gravity	9.81	(N)
Traffic loads on to the ground	55.4	(Kpa)

Table 5.
The type of loads applied in the numerical model.

gravitational loads, while the load is represented Traffic in general, with a live load as a pressure load on the surface backfill [15], and its value is estimated from the vehicle load on the surface trench surface area,

$$\text{Traffic load} = \frac{72\ \text{KN}}{(0.5 + 0.3 + 0.5) * 1\ \text{m}^2} = 55.4\ \text{KN/m}^2 \tag{1}$$

Loads have been applied to the model in one step. This step has been called static general step. Implicit code (standard) has been used to find the solution to the model.

7. Simulation results

In this paper, the results of numerical calculations of deformation pipe were presented. Four different types of surrounding soils were considered. More specific conclusions are following:

7.1 Effect of backfill depth on the maximum crown bending moment of buried concrete sewerage pipeline

Figure 10 shows the effect of the backfill height on the maximum crown bending moment of buried concrete sewerage pipeline and for all of the backfill conditions considered if it was compacted or uncompacted backfill soil. It can be seen that increasing the backfill height linearly increases the maximum crown bending moment of buried concrete sewerage pipeline for all of the backfill types.

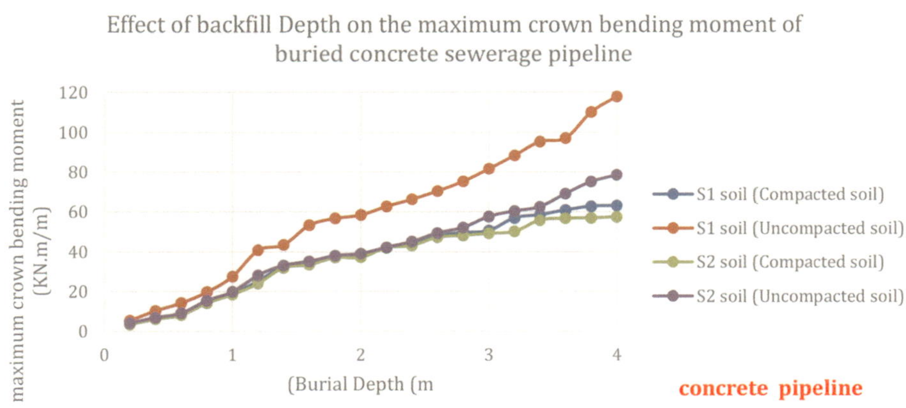

Figure 10.
Effect of backfill depth on the maximum crown bending moment of buried concrete sewerage pipeline.

Figures 11 and **12** shows the effect of the backfill types on the maximum crown bending moment concrete pipeline element developed around the pipeline. It can be seen that the maximum crown bending moment at the invert and the crown of the concrete pipeline element increases as the backfill quality decreases, for example changing the backfill type from compacted soil (S2) to uncompacted soil (S1) increases the bending moment by 79%. This is due concentration of the reaction forces at the invert and the crown of the concrete pipeline as the quality of the soil in the haunch and pleura zone decreases. However, it can be seen that there is no significant increase in the bending moment at the invert and the crown of the concrete pipeline element as the backfill type changes from compacted soil (S2) to compacted soil (S1),

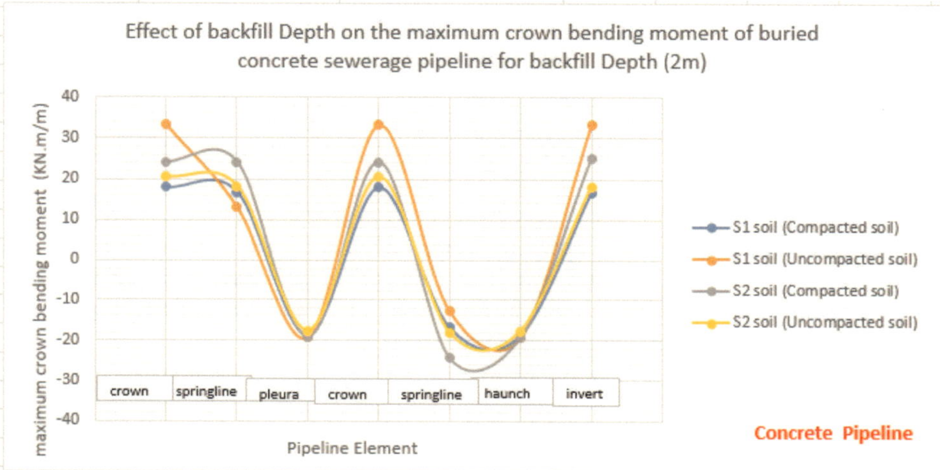

Figure 11.
Effect of backfill depth on the maximum crown bending moment of buried concrete sewerage pipeline for backfill depth (2 m).

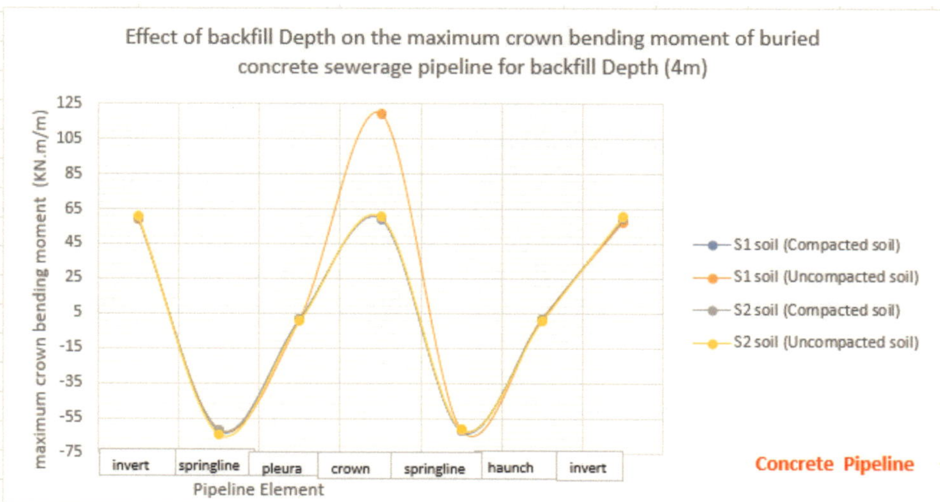

Figure 12.
Effect of backfill depth on the maximum crown bending moment of buried concrete sewerage pipeline for backfill depth (4 m).

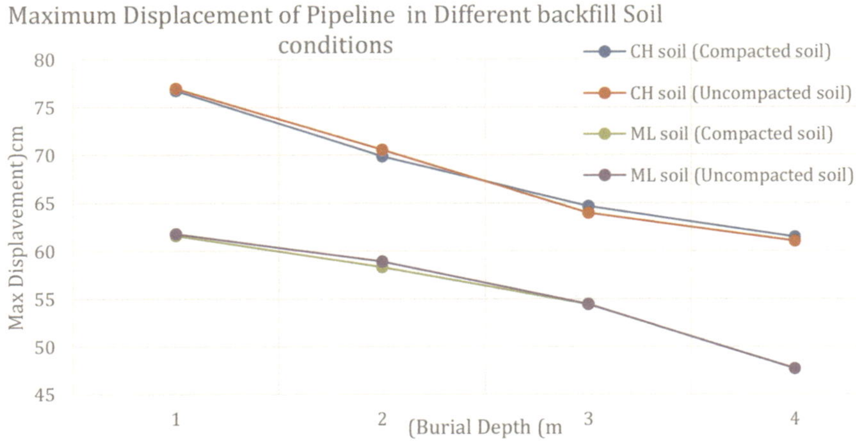

Figure 13.
Maximum displacement of buried concrete sewerage pipeline in different backfill soil conditions.

where the percentage increase is equal to 5.2% and 8.1% for the 2 m and 4 m backfill soil, respectively. The figure also shows that the backfill type does not significantly affect the bending moment developed at the crown and the invert of the concrete pipeline (the maximum percentage difference is 22%).

7.2 Displacement of the soil

Present study has been represented the immediate settlement of (ML, CH) backfill soil under loads conditions for both compacted backfill soil and uncompacted backfill soil.

Figure 13 describes the maximum displacement occurred in (CH) soil for both compacted and uncompacted soil. It has been observed from the plot that immediate maximum displacement of uncompacted soil is greater for the compacted condition than the ML soil condition. Soil settlement due to the loads have been affected by the vertical load only and maximum settlement has been found on the vertical direction (U2).

Also figure describes the maximum displacement of (CH) soil for both compacted and uncompacted soil, where it can be found the displacement due load in uncompacted soil is also greater than the compacted condition. In uncompacted condition, (CH) soil has been showed larger displacement than the (ML) soil. In compacted condition, (CH) soil has showed more displacement than the (ML) soil. **Figure 13** explains the relative displacement for different backfill soil conditions.

8. Conclusions

Present study has been carried out to investigate the failure criteria of soil and pipelines in different types of backfill soil under loads. The finite element method has been used in the platform of ABAQUS to develop the soil-pipeline model. Displacements have been investigated for different type of backfill soil in (ML, CH) under loads. The following conclusions can be drawn from the study:

1. Backfill soil properties are necessary to study the mechanical interaction between soil and buried concrete sewerage pipelines by simulating the stress and strain behavior in soil and determining the effect of backfill soil parameters. The interactions between the pipeline and the soil are greatly affected by the properties of the backfill soil and the method of laying the buried pipeline in the trench just like in compacted soil.

2. The results showed that the backfill condition affects the maximum bending moment developed at the crown and the invert of the concrete pipeline, where the maximum percentage difference was 22%.

3. The maximum crown bending moment in the buried concrete sewerage pipelines, increased by 79%, as the backfill compacted soil (S2) to uncompacted soil (S1).

4. Changing the backfill type from compacted soil (S2) to compacted soil (S1) does not significantly affect the maximum crown bending moment for the buried concrete sewerage pipelines.

5. The maximum displacement of (CH, ML) soil for uncompacted soil is greater than the compacted condition.

6. The uncompacted condition backfills (CH) has been showed larger displacement than the (ML) backfill. In compacted condition, (CH) soil has showed more displacement than the (ML) soil.

Author details

Nariman Hisham Halalo
Geotechnical Engineering, Department of Geotechnical Engineering, Civil Engineering Faculty, Damascus University, Damascus, Syrian Arab Republic

*Address all correspondence to: nar-119@hotmail.com

References

[1] Alani AM, Faramarzi A, Mahmoodian M, Tee KF. Prediction of supplied build-up in filled sewerage pipes. Environmental Technology. 2014; **35**:1721-1728

[2] American Society of Civil Engineers (ASCE). Guidelines for the seismic design of oil and gas pipeline systems. In: ASCE Committee on Gas and Liquid Fuel Lifelines of the ASCE Technical Council on Lifeline Earthquake Engineering. New York, USA. 1984

[3] Nariman H. Undrained shear behavior of samples disturbance of unsaturated expansive clay soils. Arab Journal of Sciences & Research Publishing, National Research Center-Gaza-Palestine. 2021;**5**(1):7-8. DOI: 10.26389/AJSRP.L120820

[4] Zhao Q, Kuraoka S, Baker THW, Masson P, Gu J-F, Boudreau S, et al. Durability and Performance of Gravity Pipes: A State-of-the-Art Literature Review. Ottawa, Canada: National Research Council of Canada, Urban Infrastructure Rehabilitation Program; 1998

[5] Howard A, K. Pipe Bedding and Backfill, United States Department of the Interior, Bureau of Reclamation, Technical Service Center, Geotechnical Services. 2nd ed. Denver, Colorado: Geotechnical Training Manual Np. 7; 1996

[6] Watkins, Keil, Mielke, Rahman. Pipe Zone Bedding and Backfill: A Flexible Pipe Perspective. In: Pipelines 2010: Climbing New Peaks to Infrastructure Reliability—Renew, Rehab, and Reinvest. 2010

[7] Terzi NU, Yılmazturk F, Yıldırım S, Kilic H. Experimental investigations of backfill conditions on the performance of high-density polyethelene pipes. Experimental Techniques. 2012;**36**: 40-49

[8] Brachman RWI, Moore ID, Munro SM. Compaction effects on strains within profiled thermoplastic pipes. Geosynthetics International. 2008; **15**:72-85

[9] ASTM D2321. Standard Practice for Underground Installation of Thermoplastic Pipe for Sewerages and Other Gravity-Flow Applications. USA: American National Standard; 2000

[10] ABAQUS. "ABAQUS Documentation", Dassault Systèmes, Providence, RI, USA. BSI. "Nonalloy steel tubes and fittings for the conveyance of water and other aqueous liquid-technical delivery conditions". London: BS 10224:2002; 2011

[11] AASHTO. American Association of State Highway and Transportation Officials. 3rd ed. Washington, DC: AASHTO; 1994

[12] AWWA Manual M9. Concrete Pressure Pipe. 3rd ed. Denver CO: American Water Works Association; 2008

[13] AWWA Manual M23. PVC Pipe – Design and Installation. 3rd ed. Denver CO: American Water Works Association; 2020

[14] AWWA Manual M55. PE Pipe – Design and Installation. 2nd ed. Denver CO: American Water Works Association; 2020

[15] Rao SS. The Finite Element Method in Engineering. Boston: Butterworth Heinemann; 1999

[16] Liu GS, Li L, Yang XC, Guo LJ. A numerical analysis of the stress distribution in backfilled stopes considering nonplanar interfaces between the backfill and rock walls. International Journal of Geotechnical Engineering. 2016;**10**:271-282

Chapter 5

Nanohybrid Polymer Concrete: Advancing towards Sustainable and Durable Infrastructure

Abanoub M. AbdElmaseih

Abstract

Polymer concrete is a modern construction material becoming increasingly popular as a greener alternative to conventional cement-based concrete. Using polymer binders instead of cement in the concrete mixture helps reduce the environmental impact and enhance the material's properties. This proposal aims to focus on utilizing synthetic polymer resin as the primary binder in the polymer concrete mixture. This type of resin is known for its excellent binding properties and ability to improve the overall strength and durability of the concrete. Furthermore, we plan to integrate nanotechnology further to enhance the polymer concrete's performance and sustainability. Adding nanomaterials, such as metal nanoparticles, can help improve the concrete's mechanical properties, increase its resistance to corrosion and wear, and reduce its overall weight and carbon footprint. Overall, this proposal presents an innovative and eco-friendly approach to concrete construction, potentially revolutionizing the industry and promoting sustainable development.

Keywords: polymer concrete, sustainability, synthetic polymer resin, nanotechnology, mechanical properties

1. Introduction

Due to the nature of the monomers and the curing process, polymer concrete (PC) has less environmental impact and saves energy and resources. However, the addition of a large amount of monomers in the mix may raise the cost of materials, so the monomer used is very important, as is the filler added to the mix. Interestingly, the use of nanomaterials in the construction sector is good for some properties, mechanical and durability, so their use in monomer concrete must prove to be of great technological and economic interest and beneficial due to their high potential effect. This study consolidates the world's opinion about the incorporation of nanoparticles into polymer concrete.

In response to growing environmental and resource conservation problems, today's plastics still lack attention-grabbing alternatives, including the development of more affordable products. It is important to note the contribution that concrete has to the mentioned problems because it is one of the materials most used in construction. In turn, when added to concrete, polymers improve different physicochemical properties of ordinary concrete and materials with technical, economic, and environmental aspects, which are called polymer concrete.

1.1 Background and significance

The current reality poses tremendous challenges for modern society. These challenges include rapid growth in world consumerism, unprofessional attitudes about waste disposal and recycling, and a constant lack of resources. The recombination and recycling of waste or by-products from the mining of various rocks, the machining of quartz-bearing rocks, and the production of high-technology equipment have become key strategies for achieving a sustainable global economy. At the same time, it is necessary to recognize and explore advanced materials with properties that convey higher durability and strength. The addition of various additives, dispersions, foam fibers, or composites to conventional concrete can significantly improve its engineering properties. Among these, polymer concrete contributes to obtaining an increased strength and modulus of elasticity, a decreased porosity, and improved resistance to temperature, acids, chloride ions, and sulfate ions. On the other hand, the incorporation of nanoparticles can provide the expected reinforcing effect, especially from metal nanomaterials, and can stimulate the completion of polymerization reactions, involving the polymeric matrix, leading to polymer concrete with special engineering characteristics.

1.2 Research objectives

At the same time, there is a tandem: polymer concrete is increasingly being used in structures in an environment in which the degradation of organic binders is becoming increasingly important. Polymers have been widely used in civil engineering due to their good ductility and mechanical properties, targeted formation of the microstructure, and consumer properties, the product of which also depends on the qualitative composition of the components. The major drawbacks of the majority of polymer materials used in civil engineering are poor fire resistance and low thermal conductivity. In filling such constructions in the event of a fire, polymer concrete flow is non-stationary at elevated temperatures due to volatile release and lack of mass and energy balance in the system. The softening and swelling of the polymer matrix in the temperature range of 200–700°C can significantly degrade the characteristics of the material. Given the high temperature, the study of the properties of polymer composite materials with incorporated nanoparticles at elevated temperatures is becoming increasingly relevant.

2. Fundamentals of polymer concrete

There is a constant need for construction materials with new and enhanced performance features due to advancements in civil engineering and industrial expansion. As a result of the growing environmental pollution and the necessity to preserve and extend the lifespan of building structures, modern building materials must not only possess strong properties but also exhibit chemical resistance. Innovative and cutting-edge, polymer concrete meets all the rigorous standards for chemical resistance and durability while providing excellent mechanical strength [1].

A composite material known as polymer concrete (PC) has a binder made completely of synthetic organic polymers. Other names for it include plastic resin concrete, synthetic resin concrete, and just resin concrete. Polymers should only be used in applications where the greater cost can be justified by improved characteristics, low labor costs, or low energy needs during manufacturing and handling since using a polymer instead of Portland cement represents a significant increase in cost. Therefore, to choose the best and most affordable product for a given application, architects and engineers need to have some understanding of the capabilities and limits of PC materials [2].

Since conditions frequently demand particular material needs that may be addressed by PC when numerous composite qualities are taken into consideration at the same time, polymer concrete (PC) is shown to be the most suited material for the intended application in many cases.

Because of its quick curing time, superior strength and durability, superior damping capabilities, and plenty of elastic moduli, PC is an incredibly adaptable material with a wide range of uses. Its main drawbacks are its expensive cost (the cost of the binder can vary from less than $1 to many dollars per pound), the characteristics' sensitivity to temperature, the flammability and volatility of the monomers and resins, and the fact that many users lack prior PC knowledge [3, 4].

Comprehending the characteristics of PC is essential for creating materials with the appropriate qualities and designing the most economical PC composites. Although the polymer industry has advanced significantly over the previous 50 years, it is expected that new and enhanced polymers will be created [3], such as:

- Polymers with more stable properties over a wide range of temperatures.
- A much wider range of polymers that are compatible with fresh concrete.
- Monomers, perhaps in the form of vapors, can be used for producing PC much more rapidly and simply.
- Resins that are designed to be recycled.

2.1 Composition and properties

The components of polymer concrete are a polymer binder, which is often a thermosetting polymer but can also be a thermoplastic, and a mineral filler, such as aggregate (**Figure 1**).

The composite material is known as a polymer mortar when sand is utilized as a filler. Crushed stone, gravel, limestone, chalk, expanded glass, granite, quartz, clay, condensed silica fume (silica flour, silica dust), and metallic fillers are some more fillers. Generally speaking, filler can be any solid, dry, and nonabsorbent substance [4].

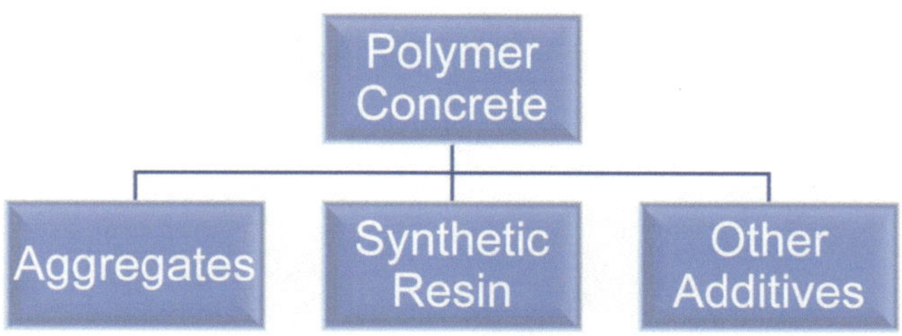

Figure 1.
Composition of a polymer concrete.

A composite material called polymer concrete (PC) is created by bonding aggregates together in a polymer matrix using resins. Different kinds and combined amounts of materials and resins have a significant impact on PC performance [5].

Highly filled polymer compositions, such as polymer concrete, may be created on any synthetic binder. However, around 10 distinct types of monomers or oligomers are employed in practice because of the need for density, strength, deformability, chemical resistance, and other qualities. They produce more than 30 types of polymer concrete when combined with modifying additives.

The kind of binder used in the polymer concretes—furan, polyester, epoxy, phenol formaldehyde, carbamide, and so forth—distinguishes them from one another. **Figure 2** illustrates how the primary varieties of polymer concrete are categorized based on the kind of synthetic resins utilized [6].

A range of aggregate materials, including fine fly ash, phosphor-gypsum, cinder, silica fume, limestone, granite, quartz, clay, crushed stone, silica sand, or calcium carbonate ($CaCO_3$), has been used in PC silicates. The foundry sector has employed a variety of silica sands [7]. Aggregates used must be usually dry and free of dirt to get the best bond between aggregates and resin.

To achieve the lowest feasible void volume, the PC mix design usually employs an aggregate size gradation and uses the least amount of polymeric binder required to cover the aggregates. PC is one such mixture, consisting of 85-weight percent siliceous sand with particle diameters ranging from 245 to 342 mm and 15-weight percent resin.

A monomer or prepolymer—a product obtained from the partial polymerization of a monomer—a catalyst, a hardener (cross-linking agent), and a filler are combined to create PC. Fire retardants and plasticizers are additional components that are added to the mixture. Silane coupling agents are occasionally employed to strengthen the binding between the filler and the polymer matrix. It is easy to change the duration required for the formation of maximal strength, ranging from a few minutes to many hours, by modifying the catalyst system and temperature. The size of the filler usually dictates how much polymer binder is required, which is often rather small. Normally the polymer content will range from 5% to 15% of the total weight, but if the filler is fine, up to 30% may be required [2, 8].

Various fiber reinforcements are utilized in polymer concrete products to help them reach their maximum potential for certain applications. The PC matrix can be supplemented with natural or synthetic fibers, such as carbon or glass fibers, to enhance the mechanical properties of polymer concretes. The kind of fiber and how much of it is in the PC determine how much mechanical improvement there is. For

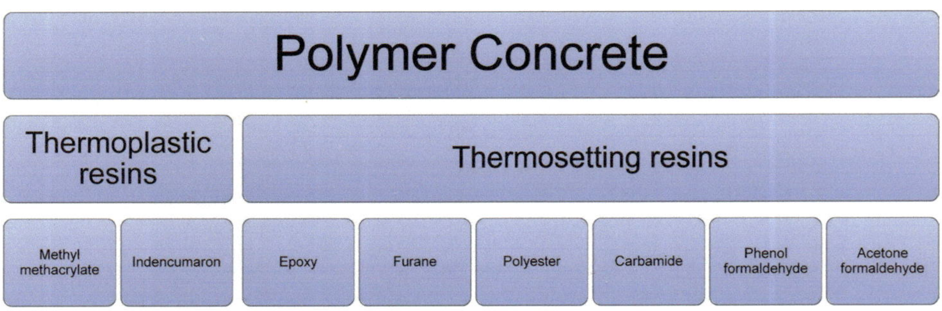

Figure 2.
Classification of polymer concretes.

instance, glass and organic fibers significantly improve the post-cracking response, which results in increased tensile, flexural, and impact strength in addition to better toughness and ductility. However, they have minimal influence on the pre-cracking behavior. Examples of the most popular fiber reinforcements are as follows [7, 9]:

- Polyester fibers: An improvement of 95% in flexural strength has been obtained when adding 2% glass fibers and using silane as a coupling agent.

- Polyacrylonitrile (PAN) carbon fibers up to 6% (w/w) were used to reinforce PC. In compression, it increased the failure strain, but the strength and modulus decreased. In tension, the addition of carbon fibers increased failure strain, strength, and modulus [9].

- Glass fibers are noncorrosive, nonconductive, and nonmagnetic, and offer low density and high modulus. The addition of glass fibers enhanced flexural strength and toughness.

- Fracture properties: Treatment with glass fibers further enhanced the flexural properties of PC.

- Different steel fibers were used; the effect of steel fibers in a PC system manifests as an increase in compressive, flexural, and impact strengths.

Adding orientated fibers is a specific type of fiber reinforcement. Placed along the primary stress directions, reinforcement glass fibers and plastic bars lessen creep deformation, which, if prevalent to a significant extent, might lead to structural collapse or impairment.

2.2 Types of polymer concretes

Polymer concrete is made using a broad range of prepolymers and monomers. Based on five different types of prepolymer systems, or monomers, the most often used polymers are:

- Epoxy

- Carbamide (urea-formaldehyde)

- Acryl (methyl methacrylate)

- Furan (furfuryl alcohol)

- Polyester

In this section of the chapter, we'll examine the range of qualities typically found in polymer concrete products made with each of these polymers.

2.2.1 Epoxy polymer concrete

Thermosetting polymers are used in epoxy binders. Several curing agents may be employed to harden epoxy polymers, but polyamines (e.g., tertiary polyamines) are the most often utilized. PC goods with the highest level of chemical resistance are produced by using polyamine hardeners, also known as curing agents. Polysulfide polymers and polyamides are additional curing agents. Epoxy PC goods cured with polyamides are more flexible, more heat resistant, and less likely to chalk when exposed to the outdoors; but they are less solvent and chemical-resistant than comparable products cured with polyamines. Polysulfide polymers are added to epoxy PC to increase its flexibility.

The process of making epoxy polymer concrete involves the use of crushed aggregates with varying diameters (0.4 and 0.8 mm), epoxy resin, and silica fume as filler [10].

Epoxy PC has excellent adhesive strength to a wide range of substrates, minimal shrinkage during setting and post setting, strong resistance to acids and alkalis (but not durable in oxidizing media), and good fatigue and creep resistance.

Epoxy polymers have not been utilized as binders in PC goods very often due to their high cost. Since the higher cost can be justified in certain cases, epoxy PC is only used for specialized applications. Examples of these include skid-resistant overlays (filled with sand, emery, pumice, and quartz) on highways, epoxy plaster for exterior walls (such as exposed aggregate panels), and resurfacing material for deteriorated areas (such as flooring).

Transparent panels, boat hulls, and vehicle bodywork are all made of epoxy polymer reinforced with glass, carbon, or boron fibers [2].

Epoxy resins are commonly used in a variety of applications due to their excellent adhesive and mechanical properties. However, these resins tend to be highly viscous and have a short pot life, which makes them difficult to work with. To address these issues, solvents are often added to the resin to dilute it. The addition of solvents not only thins out the resin but also extends its pot life and reduces its viscosity. This makes it easier to mix, apply, and work with the resin, improving its overall performance. However, when using volatile solvents, such as acetone or toluene, there is a risk of the binder developing holes and capillaries. This can occur when the solvents evaporate and leave behind voids in the polymer. As a result, the density of the polymer is reduced, which can impact its mechanical properties. To mitigate this issue, lower-volatility solvents can be used, such as alcohols or glycols, which have a slower evaporation rate and are less likely to cause voids in the polymer. It's important to consider the specific requirements of the application when selecting the appropriate solvent to use [6].

Studies have shown that the physical strength of manufactured epoxy PCs is significantly influenced by the amount of resin and filler present in their chemical composition. The ideal composition for maximal compressive and flexural strength is a PC sample consisting of 200% filler and 15% resin. On the other hand, for maximizing tensile strength, the optimal composition is 200% filler and 20% resin. It is noteworthy that prefabricated PCs exhibit mechanical strength that is four to five times higher than that of Portland cement concrete, making them an excellent choice for applications where high strength is required. With such exceptional strength properties, epoxy PCs are widely used in the construction, aerospace, automotive, and marine industries [5].

2.2.2 Carbamide polymer concrete

The polycondensation reaction of urea and formaldehyde in an aqueous or aqueous-alcoholic solution yields carbamide (urea-formaldehyde) resin. Polymer concretes made with carbamide resins are advantageous to manufacture and have a low toxicity level. Comparatively speaking to other kinds of polymer concretes, these PCs have lower mechanical and physical qualities and a higher (up to 30%) polymer matrix content.

It should be mentioned that 30–40% of the resin in carbamide resins is tree water, which causes significant shrinkage when the composition cures. This can occasionally result in the material breaking [6].

2.2.3 Acryl polymer concrete

Methacrylate (MMA) may be polymerized to produce polymethyl methacrylate, the most widely used acrylic polymer. The substance is made up of a matrix created by cross-linking MMA with polyfunctional acrylic oligomers, such as trimethylolpropane trimethacrylate, and a highway-grade aggregate.

With its excellent weathering resistance, good waterproofing qualities, good chemical resistance, and relatively low setting shrinkage (0.01–0.1%), PC made with this acrylic polymer as a binder is a versatile material whose coefficient of thermal expansion is comparable to that of Portland cement concrete.

Acryl PC has an extremely high freeze-thaw resistance due to its extremely low inclination to absorb water. All the same, the MMA monomer's low flash point (11°C) is a drawback since it poses a safety risk.

Despite being more costly than the prepolymer monomer used in the more widely used polyester PC, the MMA monomer's special qualities allow for its usage in a wide range of different applications, such as the production of curbstones, stair units, sanitary items, and facades.

The material's capacity to withstand low temperatures and its long-term endurance make it perfect for urgent repairs when conventional concrete repair materials aren't suitable.

Its application as a fast-curing structural patching material for large-hole bridge deck repairs, spall repairs, emergency full-depth repairs, bridge expansion joint headers, bearing pads, closure pours, and concrete structural beam repairs has proven to be a very successful product.

As soon as the primer has dried, the MMA polymer concrete may be placed at any moment, allowing for the effective placement of various places by prepping them beforehand [11].

2.2.4 Furan polymer concrete

Furan resins are among the several binders that have the best resistance to heat, bases, strong acids, organic solvents, etc. Therefore, compared to other synthetic resin-based binder systems such as unsaturated polyester, epoxy, furan resins provide several benefits when properly cured. In the chemical industry and other sectors where corrosive conditions are present, such as power plants, steel mills, and the paper industry, furan-based laminating resins reinforced with glass fibers are widely utilized as building materials. Reaction containers, floors, pits, trenches, catch basins, and other locations are only a few of the numerous applications for the items [12].

Furan polymers are created from furfuryl alcohol, a substance obtained from agricultural waste like corncobs, rice hulls, oat hulls, or sugar cane bagasse. The furan prepolymer is commonly combined with furfuryl alcohol, furfuraldehyde, or formaldehyde to form thermosetting polymers that are highly resistant to various aqueous acidic or basic solutions as well as strong solvents, such as ketones, aromatics, and chlorinated compounds. One significant feature of furan resins is their ability to be stored for extended periods, even up to 5 years, even at low temperatures. As such, they are a staple supply and reasonably priced when compared to other synthetic resins. More often utilized are the condensation products of furfuraldehyde and acetone at various mole ratios [13].

Even though other kinds of furan resins, including a copolymer of furfuryl alcohol and furfuraldehyde or a homopolymer of furfuryl alcohol, are also known to be effective binders for applications requiring chemical resistance. Despite this, there aren't many references in this area in literature [12, 13]. Furan resin is therefore a great option for polymer concrete applications because of its well-balanced properties. Nevertheless, these resins' short pot life and toxicity while uncured make them difficult to utilize, and high self-heating temperatures lead to considerable thermal stresses that negatively impact furan PC's strength properties. The mechanical characteristics of the composite were seen to diminish as a result of the aging-induced breakdown of furan polymers.

In order to create chemically resistant brick floors (such as carbon and red shale brick) and linings, furan polymers are utilized as binders in mortars and grouts. These flooring not only offer exceptional resistance to chemicals, but they also show outstanding resilience to high temperatures and severe thermal stress [11].

2.2.5 Polyester polymer concrete

It has been more than 20 years since polyester concrete was employed in buildings. When rapid material consolidation is required, this kind of polymer concrete can effectively replace conventional Portland concrete. Excellent chemical resistance and physical characteristics, such as high strength and little shrinkage, are displayed by polyester polymer concrete building material.

A form of thermosetting resin produced via polycondensation is polyester resin, such as epoxy. Low viscosity and excellent mechanical and electrical insulating qualities, together with a strong resistance to acids, oils, and petrol, characterize products based on them. Unsaturated polyester resin (UP), along with a certain amount of filler, fine aggregate, and coarse aggregate, are the ingredients of polyester polymer concrete, a synthetic material. In contrast to OPC, PPC employs unsaturated polyester resin as a concrete binder rather than cement.

Given their affordability, accessibility, and superior mechanical qualities, unsaturated polyester resins are the most widely utilized resin systems in polymer concrete [14].

The US Patent 4371639A described an innovation for curable polymer concrete compositions that are especially appropriate for construction-related items. These compositions consist of three components: (1) a polyester composition, (2) an aggregate mix that includes both sand and fly ash, and (3) an initiator [15]. These mixes are particularly well suited for usage in constructing panels, piping, bridge decks, spillways, and tank linings.

The construction performance and mechanical qualities of polyester PC will be directly impacted by the unsaturated resin curing process. In contrast to OPC, PPC frequently exhibits high viscosity during stirring when resins are present, requires much faster stirring needs, and requires forceful agitation throughout the mixing process.

When it comes to liquid resin and curing agent, initiator, accelerator type, and dose, PPC's hardening process differs from that of OPC and asphalt concrete [6].

Unsaturated polyester polymers are widely used in various industrial and commercial applications due to their exceptional physical and chemical properties. These polymers are usually produced by reacting unsaturated dibasic acids or anhydrides with diols or polyols. The resulting prepolymer is then dissolved in co-polymerizable monomers, such as styrene or a combination of styrene and methyl methacrylate, to form a solution. This solution typically contains 60–80% prepolymer, depending on the intended application. The use of co-polymerizable monomers in the solution has several advantages. First, they help to reduce the viscosity of the prepolymer, making it easier to handle and process. Second, they improve the mechanical and thermal properties of the resulting composite material. Third, they provide a cost-effective way to produce unsaturated polyester polymers, making them more accessible to a wider range of industries and applications.

Polyester resins should be noted for their toxicity. Additionally, stratification and level-by-level curing of the composition are likely to occur if the mixture's components are not stirred well.

Polyester PC has excellent chemical and freeze-thaw resistance along with great mechanical, cohesive, and adhesive strength; however, its significant setting and post-setting shrinkage, which can reach up to 10 times more than Portland cement concrete, imposes a significant drawback in some situations.

Polyester polymer-based concrete is highly resistant to a variety of harmful substances, such as oxidizers, acids, oils, and petroleum compounds. However, it is important to note that polyester polymer-based concrete has some limitations. It is not as resistant to alkaline solutions or water as it is to other substances. When exposed to water, the concrete tends to lose its strength faster than when it is exposed to inorganic salt or certain acid solutions. Therefore, when determining the level of water resistance of the concrete, the resistance to acid solutions is used as an indicator. This means that if the concrete is resistant to acid solutions, it is likely to be more water-resistant than if it is not [16].

2.2.6 Other types of polymer concrete

The use of methyl methacrylate, acetone formaldehyde resins and monomers, and phenol-formaldehyde in polymer concretes is far less prevalent. In several mechanical and physical aspects, phenolic resins resemble furans. Nevertheless, in alkalis such as polyester resins, they exhibit volatility [6].

Phenol-formaldehyde resins bear several physical and mechanical similarities to furan, among them. They resist heat and fire well, have minimal permeability to liquids, and have strong compressive and flexural strengths. However, because they tend to hydrolyze and lose their potency in alkaline settings, they are not recommended for use in certain conditions.

Conversely, acetone formaldehyde resins and monomers are less often utilized in polymer concretes because of their high price and scarce supply. Though not as strong as furan or phenolic resins, they are nonetheless weatherable, have acceptable mechanical qualities, and are resistant to chemicals.

Polymer concretes, particularly those intended for flooring purposes, use methyl methacrylate as a polymer binder. It has strong abrasion resistance, outstanding color and gloss retention, and superb adherence to concrete substrates. However, heavy-duty applications or situations requiring strong impact or thermal shock resistance should not be used.

3. Metal nanoparticles in concrete

Metal nanoparticles, typically in the form of oxides or elemental metals, have unique physical and chemical properties due to their small size and high surface area-to-volume ratio. These properties can significantly influence the behavior of polymer concrete when incorporated into the matrix. For instance, metal nanoparticles, such as iron oxide (Fe_3O_4) and titanium dioxide (TiO_2), have been shown to impart enhanced mechanical strength, UV resistance, and antimicrobial properties to polymer composites. Additionally, the electrical conductivity of metal nanoparticles can contribute to the development of functional polymer concrete with applications in smart infrastructure and electromagnetic shielding [17].

3.1 Effects of metal nanoparticles on polymer concrete

The mechanical properties of polymer concrete, including compressive strength, flexural strength, and impact resistance, can be improved through the addition of metal nanoparticles. Studies have demonstrated that the dispersion of metal nanoparticles within the polymer matrix can lead to enhanced interfacial bonding and load transfer, resulting in increased strength and toughness. Furthermore, the presence of metal nanoparticles can act as reinforcing agents, effectively inhibiting crack propagation and improving the overall structural integrity of the composite material. The type, concentration, and distribution of metal nanoparticles play a crucial role in determining the extent of improvement in mechanical properties, and thus, warrant careful consideration during the design of metal nanoparticle-reinforced polymer concrete.

The thermal behavior of polymer concrete is another important aspect affected by the incorporation of metal nanoparticles. Metal nanoparticles exhibit high thermal conductivity, which can facilitate heat dissipation and improve the thermal stability of the composite material [18]. Moreover, the addition of metal nanoparticles has been shown to mitigate thermal expansion and contraction, thereby reducing the likelihood of cracking and dimensional instability at elevated temperatures. These thermal enhancements are particularly advantageous for applications in construction, aerospace, and automotive industries where exposure to varying temperatures and thermal cycling is common.

One of the most significant effects of metal nanoparticles in polymer concrete is the enhancement of electrical conductivity and electromagnetic interference (EMI) shielding effectiveness. Metal nanoparticles, such as silver (Ag) and copper (Cu), are renowned for their excellent electrical conductivity, and when uniformly dispersed in the polymer matrix, can impart conductive pathways throughout the material. This property is highly desirable for applications requiring antistatic flooring, electrical heating elements, and electromagnetic shielding in electronic devices. The ability of metal nanoparticle-reinforced polymer concrete to attenuate electromagnetic radiation makes it a promising candidate for the development of EMI shielding materials in telecommunications, aerospace, and defense sectors.

3.2 Fabrication methods and dispersion techniques

The successful integration of metal nanoparticles into polymer concrete necessitates effective dispersion and interfacial interaction between the nanoparticles and the polymer matrix [19]. Several fabrication methods and dispersion techniques have been explored to achieve homogeneous distribution and optimal bonding of metal nanoparticles within the composite. Common approaches include solution blending, melt mixing, in-situ polymerization, and surface modification of nanoparticles. Solution blending involves the dissolution of metal nanoparticles in a polymer solvent followed by evaporation to obtain a homogeneous mixture, while melt mixing entails the direct compounding of molten polymer with metal nanoparticles. In-situ polymerization methods involve the simultaneous formation of the polymer matrix and the dispersion of nanoparticles, leading to improved interfacial adhesion. Surface modification techniques, such as silane coupling agents and surfactant treatment, are employed to enhance the compatibility and dispersibility of metal nanoparticles in the polymer matrix. Each fabrication method and dispersion technique offer distinct advantages and challenges, and the selection of the most suitable approach depends on the specific requirements of the metal nanoparticle-reinforced polymer concrete.

4. Sustainable aspects of polymer concrete

The most crucial problem facing all upcoming development trends is how to increase the sustainability of concrete. According to the Brundtland Commission Report (1987), sustainability is "development that meets the needs of the present without compromising the ability of future generations to meet their own needs." Every facet of society is included in sustainability, including life cycle assessments, energy, the environment, civil infrastructure systems, health, and safety. The most often utilized material is concrete, which uses many resources. Modern concrete normally has a mass percentage of 16–20% cementitious ingredients, 6–8% water for mixing, 60–70% aggregate, and 2–3% admixtures. The most accurate method for estimating the global production of concrete is first to calculate the quantity of cement produced, then use the mass ratio of standard practice to calculate the number of other components, and lastly put them together. An estimated 2.5 billion tons of cement, 1 billion tons of mixing water, 10 billion tons of aggregates, and 400 million tons of chemical admixtures are used yearly in concrete production, based on the quantity of cement produced in 2007. As a result, the global concrete industry grows to consume 14 billion tons of natural resources annually. Massive amounts of aggregate must be mined, processed, and transported, consuming a great deal of energy and harming virgin land ecosystems in addition to the billions of tons of raw ingredients required to produce cement. In addition, a significant amount of CO_2 is generated. Such massive natural resource use undoubtedly has a negative influence on the environment and, if unchecked, would make it more difficult for future generations to fulfill their demands. Three factors may be taken into account when analyzing the sustainability of concrete: ways to cut back on energy use, petrol emissions, particularly CO_2 emissions, and ways to use fewer natural resources for raw materials during the building, application, and manufacturing of concrete. Reduction, reuse, and recycling are the three strategies for making concrete sustainable.

According to its definition, sustainable development is any process that satisfies current demands without jeopardizing the ability of future generations to satisfy their own. Conscientious management of natural resources is crucial to preserving the environment and future generations without compromising their standard of living. As a result, while considering ecological aspects of sustainability, social, cultural, and economic factors must also be taken into account. Because of environmental concerns, the development of sustainable building materials has been a major focus in recent decades. Sustainability in construction is achieved due to the following advantages of polymers [20] in polymer concrete:

- Maximize resource reuse: Waste materials, such as fly ash, rock dust, coal mine waste, glass debris, may be used to manufacture the raw material for polymer concrete. In Japan, polymer concretes and mortars based on expanded polystyrene (EPS) solutions have recently been developed as an efficient way to use waste plastic.

- Use renewable or recyclable resources (renew/recycle): The majority of polymers exhibit superior adherence to a wide range of building materials, including bricks, stones, cement concrete, and ceramics. The excellent bonding characteristics of polymeric binders enable easy renewal.

- Protect the natural environment (protect nature): Depending on the exposure circumstances, the majority of PC goods typically have a service life of more than 20 years or longer. Furthermore, PC is thought to be an extremely resilient material against harsh environmental factors, including corrosion, chemical assault, and freeze-thaw damage.

- Create a healthy, nontoxic environment (non-toxicity): Before they set, the polymeric binders used in PC are poisonous and combustible, but after they do, they pose no threat at all.

- Pursue quality in creating the built environment (quality): Environmentally friendly, concrete-polymer composites are categorized as building materials. These kinds of materials preserve the environment, prolong the life of infrastructure, and save natural resources.

Restoring existing structures rather than developing new ones is one of the key components of sustainability. During its production and construction, new construction uses a lot of energy in the form of raw materials. In addition to being sustainable, polymer concrete may be utilized as a repair material for historic buildings. Because it is a long-lasting substance, fewer resources are needed throughout the construction process. Because polymer concrete is around four times stronger than traditional Portland cement concrete, less raw material is used in its production. The properties of polymers that allow for reuse, renewal, and recycling make them more environmentally friendly building materials that promote the sustainable features of construction.

4.1 Scientific utilization of more industry waste

Using industry waste or byproducts in place of raw materials like cement and aggregates is one method to make concrete more sustainable. Supplementary cementitious materials are the general term for the industry by-products used to substitute cement (SCMs). At the moment, the most widely utilized SCMs are silica fume, fly ash, limestone powders, and blast furnace slag. With a comparatively constant composition, these SCMs are available in significant and consistent quantities [21]. As in Europe, they can be added to cement at the last grinding stage of the cement manufacturing process to cut down on the quantity of clinker utilized. In order to use less cement while producing concrete, they can also be added to the mix. The use of SCMs can lower the quantity of cement or clinker, regardless of how it is carried out. Four gigajoules of energy are needed for every ton of cement produced during the energy-intensive process of making Portland cement, or clinker. CO_2 emissions from production are also significant [21]. Calculations show that 0.8–1 tons of CO_2 are released into the environment during the production of one ton of Portland cement clinker. In cement manufacture, limestone has to be decomposed as:

$$\underset{(100)}{\text{Limestone }(CaCO_3)}(1000^{\circ}C) \rightarrow \underset{(56)}{CaO} + \underset{(44)}{CO_2} \tag{1}$$

As 620 kg of CaO are present in 1 ton of cement, $CO_2 = 620 \times 44/56 = 487$ kg. During the cement-making process, burning fuel also produces CO_2. This varies depending on how far the burning process has progressed, from 320 kg to around 450 kg. Roughly 7% of the world's CO_2 emissions are caused by the world's annual cement output. Using SCMs can thereby significantly lessen the concrete industry's negative environmental effects. Along with recycling trash from the steel and power sectors, the concrete industry also lessens the environmental effect of both by using recycled raw materials in place of virgin raw materials [22]. Additionally, these SCMs have reactive silica, which when combined with CH in concrete can produce secondary C-S-H and enhance the material's qualities. Concrete's structure gets increasingly complicated at the nano and micrometer levels as a result of the addition of SCMs. Therefore, research must be done to determine how and to what degree SCMs can alter the structure of hydration products, and therefore the macroscopic characteristics of concrete. It is important to research and provide the ideal cement replacement quantity at which concrete would benefit the most. The scientific use of existing SCMs can be accomplished in this fashion. It should be noted that these SCMs exclusively in concrete in several different nations. It is necessary to investigate additional SCM resources in order to further minimize the amount of Portland cement used in concrete. These resources ought to comprise various industrial by-products such Pb/Zn mining tailing, phosphorous slag, coal gangue, and copper slag that also contain reactive silica and/or reactive alumina. They could be able to provide outcomes that are comparable to those of SCMs that are now in use. Nevertheless, due to their complicated makeup, certain hazardous components, and unclear impact on concrete qualities, comprehensive testing is required for each new individual by-product before it can be utilized with confidence. We cannot create new resources for SCMs and support the sustainable growth of concrete unless we have a deeper comprehension of their response processes and improved methods for characterizing them.

Industrial waste that can be utilized to replace aggregates includes waste glass, demolished concrete, bottom ash, and large-size slag. **Figure 3** show crushed waste glass as a replacement of fine aggregate. Glass waste particles are typically graded from 75 to 5 mm. When adding glass to concrete, the potential for an alkali-silica reaction is the primary source of worry [23]. The interaction between alkalis in cement paste pores and specific types of aggregates is known as an alkali-silica reaction (ASR), and it causes excessive expansion of concrete sections as well as severe cracking. Cement clinker contains trace levels of potassium oxide (K_2O) and sodium oxide (Na_2O). For this reason, it is customary to describe the cement chemical analysis results in terms of the oxides Na_2O and K_2O. Moreover, the alkali content of cement

Figure 3.
Recycled waste glass as an aggregate for concrete.

is commonly represented as the corresponding proportion of Na_2O by cement mass. Since the molecular weights of Na_2O and K_2O are respectively 62 and 94, the equivalent percentage of Na_2O is calculated with the formula.

$$\%Na_2O_{eq} = \%Na_2O + 0.658 \cdot \%K_2O \quad (2)$$

Na_2O and K_2O combine to generate hydroxides in concrete, which raises the pH from 12.5 to 13.5. As Na_2O_{eq} rises, so does the concentration of these hydroxides. Under some circumstances, the silica in such very alkaline solutions can react with the alkali to generate an endless swelling gel that absorbs any free water from osmosis and expands, upsetting the concrete matrix. The concrete is internally stressed by expanding gel products, which results in the distinctive map cracking of unconstrained surfaces. Loss of structural integrity may occur from alkali-silica reaction-induced cracking.

It should be noted that an increasing amount of destroyed concrete will be created as more and more concrete structures age. Another environmental issue will arise from the deposition of building trash if it cannot be recycled and utilized again. On the other hand, recycled concrete aggregates are often weaker and more porous. Therefore, it is critical to devise a strategy to resolve the issue and guarantee the caliber of the concrete composed of recycled aggregates.

5. Conclusion

In conclusion, the development of high-performance polymer-based composites (PMCs) with better characteristics has given considerable emphasis to sustainable methods. A number of methods have been investigated to improve the properties of PMCs, including the use of nanofillers, toughening agents, polymer matrix modification, and sophisticated production processes. These methods have shown to be successful in enhancing PMCs' mechanical, thermal, and electrical characteristics, increasing the range of industries in which they may be used.

Nonetheless, it is equally crucial to take these methods' effects on the environment into account. Numerous studies have emphasized how some of these methods negatively impact the environment, increasing carbon footprints, polluting waterways, and depleting natural resources, among other things. Thus, further study is required to create environmentally friendly substitutes that are sustainable. The utilization of recycled materials, bio-based polymers, and natural fibers as reinforcing materials are a few sustainable strategies that have been investigated.

In general, the sustainable methods covered in this review have a lot of promise to improve PMC qualities and advance environmental sustainability. To improve these methods and assess how they will affect the ecosystem over time, more study is necessary. Furthermore, in order to encourage the adoption of sustainable practices and support the growth of a circular economy in the composites sector, cooperation between researchers, policymakers, and industry actors is crucial.

Author details

Abanoub M. AbdElmaseih
Faculty of Science Ain Shams University, Cairo, Egypt

*Address all correspondence to: abanoubmahrous@gstd.sci.cu.edu.eg

References

[1] Intended Use of Polymer Concrete. Systemy i Technologie. Available from: http://www.sytec.pl/en/polimerobeton-en.php#wlasciwosci-betonowzywicznych

[2] Hing LE. Application of Polymer in Concrete Construction. A report submitted in partial fulfillment of the requirements for the degree of Bachelor of Civil Engineering Faculty of Civil Engineering, Universiti Teknologi, Malaysia. 2007. Available from: http://www.efka.utm.my/thesis/IMAGES/3PSM/2007/JSB/PARTS7/leeenghingsx031322awj04d07ttt.pdf

[3] Fowler DW. Polymers in concrete: Where have we been and where are we going? In: ICPIC 2001, The Tenth International Congress in Polymer Concrete; Honolulu, HI, USA. Farmington Hills, Michigan, USA: American Concrete Institute; 2001

[4] Fowler DW. State of the art in concrete polymer materials in the USA. In: ICPIC 2007, 12th International Congress in Polymer Concrete; Chancheon, Korea. Chuncheon, Korea: Kangwon National University; 2007

[5] Golestaneh M, Amini G, Najafpour GD, Beygj MA. Evaluation of mechanical strength of epoxy polymer concrete with silica powder as filler. World Applied Sciences Journal. 2010;**9**(2):216-220

[6] Chmyhov V. Resistance of rubber concrete to action of aggressive environments (Doctoral dissertation, PhD dissertation). Voronezh: Voronezh State Architectural and Civil Engineering University; 2002 (in Russian)

[7] Martinez-Barrera G, Vigueras-Santiago E, Gencel O, Hagg Lobland HE. Polymer concretes: A description and methods for modification and improvement. Journal of Materials Education. 2011;**33**(3-2):37-52

[8] Blaga A, Beaudoin JJ. Polymer Concrete, CBD-242. November 1985. Available from: http://web.mit.edu/parmstr/Public/NRCan/CanBldgDigests/cbd242_e.html

[9] Garas VY, Vipulanandan C. Review of Polyester Polymer Concrete Properties. Available from: http://www2.egr.uh.edu/~civeb1/CIGMAT/03_poster/11.pdf

[10] Barbuta M, Diaconescu RM, Harja M. Using neural networks for prediction of properties of polymer concrete with fly ash. Journal of Materials in Civil Engineering. 2012;**24**(5):523-528

[11] Figovsky O, Beilin D. Advanced Polymer Concretes and Compounds. Boca Raton, FL, USA: CRC Press Taylor & Francis Group; 2014

[12] Iroegbu AO, Hlangothi SP. Furfuryl alcohol a versatile, eco-sustainable compound in perspective. Chemistry Africa. 2019;**2**(2):223-239

[13] Solovjov VF, Pan VM, Freyhardt HC. Anisotropic flux dynamics in single-crystalline and melt-textured $YBa_2Cu_3O_{7-\delta}$. Physical Review B. 1994;**50**(18):13724

[14] Gorninski JP, Dal Molin DC, Kazmierczak CS. Strength degradation of polymer concrete in acidic environments. Cement and Concrete Composites. 2007;**29**(8):637-645

[15] Muszynski M, Birnbaum RS, Roos BA. Glucocorticoids stimulate

the production of preprocalcitonin-derived secretory peptides by a rat medullary thyroid carcinoma cell line. The Journal of Biological Chemistry. 1983;**258**(19):11678-11683

[16] Ribeiro A. Soft Plate and Impact Tectonics. Berlin, Heidelberg, Dordrecht, New York City: Springer Science & Business Media; 2002

[17] Ulusoy U. A review of particle shape effects on material properties for various engineering applications: From macro to nanoscale. Minerals. 2023;**13**(1):91

[18] Lin Y, Jia Y, Alva G, Fang G. Review on thermal conductivity enhancement, thermal properties and applications of phase change materials in thermal energy storage. Renewable and Sustainable Energy Reviews. 2018;**82**:2730-2742

[19] Bastos G, Patiño-Barbeito F, Patiño-Cambeiro F, Armesto J. Nano-inclusions applied in cement-matrix composites: A review. Materials. 2016;**9**(12):1015

[20] Bhutta DM, Ohama DY. Status of the recent research and development of concrete-polymer composites in Japan. In: Proceedings of the International Workshop on Cement Based Materials & Civil Infrastructure (CBM-CI). Karachi, Pakistan: NED Engineering University; Dec 2007. pp. 289-294

[21] Aprianti E. A huge number of artificial waste material can be supplementary cementitious material (SCM) for concrete production—A review part II. Journal of Cleaner Production. 2017;**142**:4178-4194

[22] Pellegrino C, Faleschini F, Meyer C. Recycled materials in concrete. In: Developments in the Formulation and Reinforcement of Concrete. Amsterdam, The Netherlands: Elsevier; 2019. pp. 19-54

[23] Dong W, Li W, Tao Z. A comprehensive review on performance of cementitious and geopolymeric concretes with recycled waste glass as powder, sand or cullet. Resources, Conservation and Recycling. 2021;**172**:105664

Chapter 6

Perspective Chapter: Comprehensive and New Approximate Analysis and Design Techniques for Reinforced Concrete Structural Elements

Mosleh Tohidi and Ali Bahadori-Jahromi

Abstract

In practical structural analysis and design scenarios, various software tools are commonly utilized. However, verifying the structural analysis and design can pose a significant challenge for many designers. To address this concern, the author has developed and proposed an innovative, simple, comprehensive, and reliable approximate structural analysis and design method. These methods aim to provide designers with valuable information on the final internal forces (axial/shear force, and bending moment), vertical deflections, lateral displacement/drift of buildings under lateral force, and approximate dimensions of all structural components prior to conducting software analysis and design. The preliminary estimation of beam and column dimensions may lead to an extensive trial and error process. Therefore, this study will introduce a new and reliable approximate structural analysis and design methodology using precise analytical techniques and software evaluations. This approach aims to determine approximate internal forces, establish preliminary structural dimensions, and validate the modeling, analysis, and design processes conducted through software. The methodology presented in this chapter has been applied to the analysis and design of various projects ranging from 5 to 15 stories, which were designed by the author in their capacity as the director of Alan Consulting Engineers. In addition, this chapter presents four case studies to assess the effectiveness and dependability of the proposed methods. The findings indicate discrepancies ranging from 2 to 12%.

Keywords: reinforced concrete structural systems, approximate loading, new approximate analysis techniques, and new approximate structural design, rigid frame, shear wall, coupled shear wall, foundations, spring

1. Introduction

It is obvious that selecting materials, gravity and lateral bearing systems, floor type, and foundation types poses a significant challenge in real-world projects. To achieve an optimal design, building and technical facility design and analysis are typically conducted in two stages: phase I and phase II [1].

Phase 1. In the initial phase, the architectural plans are prepared and presented to the client, taking into consideration the land's characteristics, location, building usage, and existing buildings around the project. Simultaneously, in the structural sector, the construction technology, availability of materials, skilled workers, and designer's knowledge are taken into account to suggest the optimal materials, load bearing system, floor type, and foundation system to the employer.

In the structural section, the position of skeleton components is initially verified, and their dimensions are selected through an approximate analysis and design [1]. Approximate analysis pertains to structural analysis that employs simplifications in both modeling and loading conditions.

Phase 2. After obtaining approval for one of the proposed options in the initial phase, the complete details of the drawings are prepared for all four areas. In the structural division, a separate structural model of the load-bearing system, frame, shear wall, floors, and foundation system is created, analyzed, and designed based on the approved architectural drawings from the first phase. It is obvious that, the theoretical approach to the analysis of frames is evidently time-consuming, and the optimizations encounter various challenges; therefore, the analysis and design of the structural frame or frame with shear wall is carried out using ETABS/SAP. It is important to note that SAFE can be utilized for the analysis and design of the slabs and foundations. Additionally, it should be mentioned that ETABSv2016 is capable of slab design, although the design results are generally lower compared to safe. The overall project estimate is based on the phase II drawings and is provided to the client as part of the tender documents [1].

Phase 3. In the third phase, the employer conducts a tender to identify the eligible contractor, taking into account the plans and estimates developed in phase II. The chosen contractor, who may be supervised by consulting engineers from the public or private sector, or an engineer licensed to work in the private sector, is responsible for executing the construction project [1].

2. Phase 1: Conceptual design

In this phase, the initial action to be taken involves selecting an appropriate firm to carry out geotechnical investigations on behalf of the employer. The key outcomes of such investigations include determining the bearing capacity of the soil and identifying the soil type beneath the foundation. Given that the soil type impacts the lateral forces during seismic events and the bearing capacity directly influences the dimensions of the foundation, it is crucial to approach this step with utmost care and precision.

Construction projects commonly utilize steel and concrete as primary structural materials. Structural engineers are often tasked with determining the most suitable material for a project, weighing factors such as sustainability, cost, construction speed, technology, labor expertise, and material availability. Through careful consideration of these variables and discussions with stakeholders, the optimal material is ultimately chosen for the construction project.

After choosing the type of materials, for example, concrete, all structural systems suitable for sustaining the gravity and lateral loads, types of floors, and foundations must be examined by the structural engineer, and then the most optimal systems are selected. The only quantitative criterion that should be selected based on that type of load-bearing system is the height of the structure based on the relevant code or standard, while the rest of the criteria are qualitative and depend more on the engineer's judgment. **Figure 1** provides a general suggestion for choosing the type of load-bearing system for various heights. As can be seen from **Figure 1**, for low/medium rise buildings, three load-bearing systems: rigid frame, shear wall, or a combination of frame and shear wall, can be used. For tall buildings, the tube system is suitable (**Figure 2**) [3, 4].

Classification of load-bearing systems in terms of connection type:

- Rigid connection: Concrete and steel structures (**Figure 2a**)
- Pinned connection: Steel structures, precast concrete structures (**Figure 2b**)

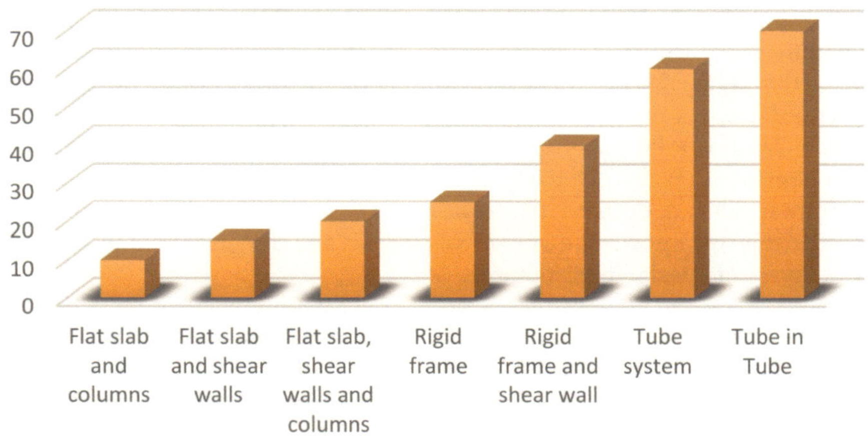

Figure 1.
Classification of structural systems based on the height of the structure.

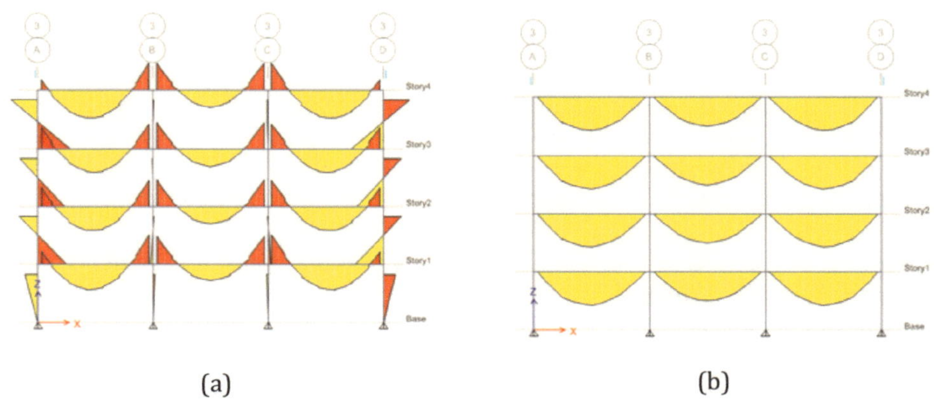

Figure 2.
Bending moment in structures with rigid and pinned connections under gravity loads using SAP 2000 [2]. (a) Rigid connections (b) Pin connections.

3. Structural analysis and design

To analyze and design a structure, the following steps need to be taken [2]:

Loading: The process of selecting and calculating the gravity and lateral loading, including wind or seismic forces, is crucial in the design of structures. It involves determining the appropriate variable, permanent, and lateral actions that act upon the structure.

Modeling (to build up computational models): To establish the geometric properties, material characteristics, cross-sectional profiles, and loading conditions and subsequently implement the determined loads onto the structural system.

Structural analysis: The impact of these actions on the structure, such as shear force, bending moment, axial forces, and deflection, will be examined through either manual or software analysis.

Verification: In order to validate a software model, it is essential to monitor the general lateral/vertical deflection, as well as the axial force/shear force/bending moment diagram under both gravity and lateral loads. These parameters must adhere to the precise specifications provided by the designer regarding deflection and internal forces, that is, axial/shear force and bending moment. To ensure the accuracy of the structural analysis results, it is necessary to employ approximate analysis techniques. If a significant disparity arises between the approximate analysis and the software-generated results, the designer must provide a clear justification; otherwise, the software results are deemed incorrect. It can be inferred that only engineers possessing extensive knowledge of manual analysis and design should utilize software for the analysis and design of structures.

Structural design and optimization: The detailed design involves conducting structural calculations to ascertain the dimensions, configuration, and specifications of structural elements and foundations, including reinforcement calculations where necessary. Adherence to the applicable Eurocodes for concrete, steel, timber, and masonry, as well as other relevant guidance documents, is essential. The design process encompasses a thorough examination of all elements to ensure accuracy and precision:

- Column, beam, and shear walls
- Beam-column connection
- Floor, that is, one-way, two-way slab, flat slab, and …
- Foundation, that is, mat, strip, pile or …
- Applying the ductility provisions for a medium or special frame

Drawing: Full details of Foundation, Columns, beams, connections, and slabs. *Preparation of tender drawings.*

Design risk assessment

Planning construction Identification of suitable forms of contract. Construction Programming and resource requirements. Outline cost analysis.

This chapter extensively covers the topics of loading and approximate analysis, which are considered powerful techniques for verifying software analysis and design.

3.1 Loading

After choosing a suitable structural system, the first step is to calculate the various loads on the structure. The minimum loads that should be considered in the design of structural systems are dead, live, snow, earthquake, and wind loads. Due to the complexity of wind and earthquake loads, in this section, only dead and live loads and a summary of wind and seismic are discussed. EC1 provides all the required details to calculate various actions, that is, dead/live (EC1-1), snow (EC1-3), and wind load (EC1-4) on the structure [5]. For seismic load, EC8 is used.

3.1.1 Dead load (permanent action)

Dead load refers to the weight of the building's fixed additions, including floors/roofs, structural weight, internal and peripheral walls, staircases, finishes, and the weight of mechanical and electrical installations. To determine the weight of each component, the subsequent steps must be followed:

1. Providing a precise details of the elements (**Figures 3-6**)

2. Determining the unit mass of various materials given in codes (**Table 1**)

3. The amount of dead load is calculated as

$$DL = \sum t_i \gamma_i \tag{1}$$

where.
t_i is the thickness of materials.
γ_i is the density of the materials.
Density of the materials.

The density of all the construction materials is given in Annex A. For example, the density of concert and mortar is given in **Table 1**:

Figures 3-6 illustrate the details and the methodology for determining the dead load of the solid slab, block-joist floor, staircases, and wall.

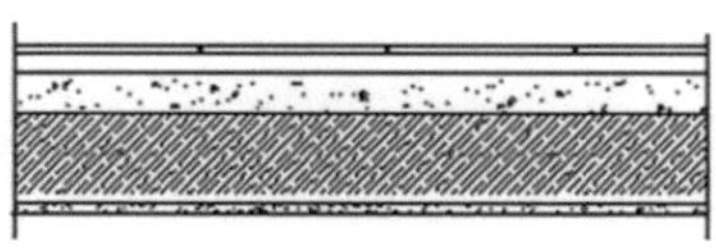

Marble	0.025*22= 0.55	kN/m^2
Sand-Cement Mortar	0.03* 21 = 0.63	kN/m^2
Light Concrete	0.1* 12 = 1.2	kN/m^2
Slab	0.15* 25 = 3.75	kN/m^2
Plaster	0.01*13=0.13	kN/m^2
Service	0.8	kN/m^2
	DL = 7.06	kN / m2
	DL=Slab+3.5kN / m2	

Figure 3.
Dead load of solid slab [2].

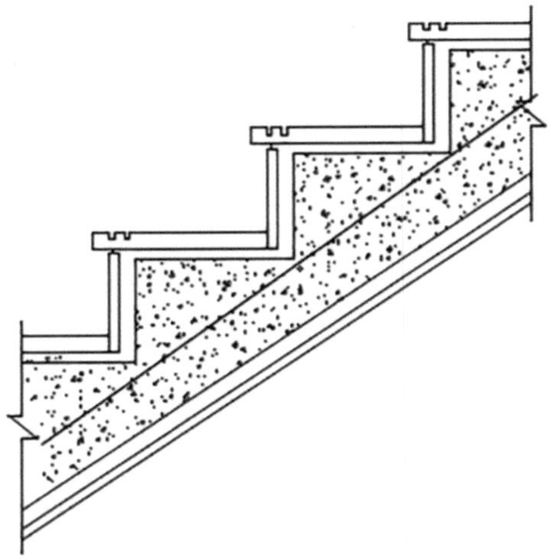

$$l = 5000\ mm\ ;\ h \geq \frac{l}{20} = 25\ cm$$

Step Bearing	0.03*27= 0.81	kN / m^2
	0.02*0.2*3.3*25 = 0.33	kN / m^2
Light Concrete	0.2*0.3*3.3*22/2/ cos(33.7°) = 2.62	kN / m^2
Slab L/20=250mm	0.25*25/cos(33.7°) = 7.5	kN / m^2
Plaster	0.02* 1600/cos(33.7°) =0. 385	kN / m^2
	DL=11.65	kN / m^2

Figure 4.
Dead load of staircase [2].

3.1.2 Live (imposed) load

During the utilization of buildings or other structures, a temporary load is imposed. An illustration of this would be the weight of books in a library, students, chairs in classrooms, or equipment found in hospitals and factories. The classification of these loads is determined by the function of the spaces in EC1. The code suggests live loads in the form of UDL, concentrated load, or line load. Typically, the structural design is established based on the UDL load and is verified for point and line loads at key locations.

EC1 uses various categories to identify the variable actions (live loads). The main categories can be listed as follows:

- Residential, social, commercial, and administration: four categories (A, B, C and D)

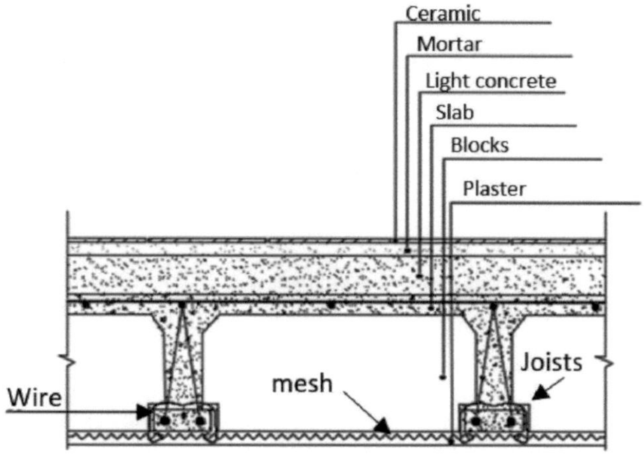

Ceramic	$0.005(22) = 0.11$	
Mortar	$0.02(21) = 0.42$	
Light concrete	$0.11(13) = 1.43$	
Slab	$0.05(25) = 1.25$	
Joists	$[0.1(0.25)/0.6](25) = 1.04$	
Plaster	$0.04(16) = 0.64$	
Whitening	$0.005(13) = 0.065$	
Services	0.2	
DL=	**5.155**	$k\,N\,m$

Figure 5.
Dead load of block and joist floor [2].

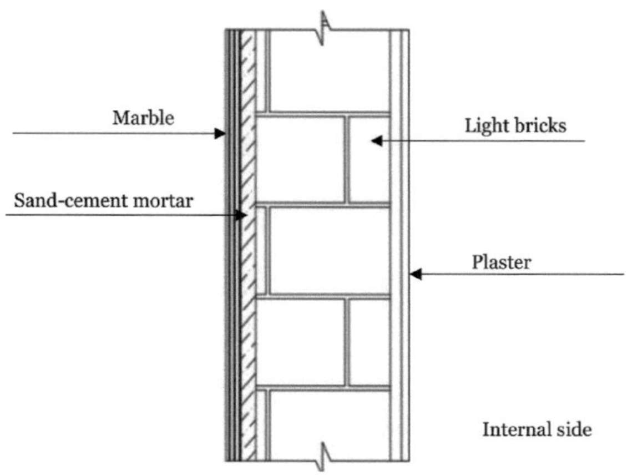

Figure 6.
Dead load of the external wall [2].

Material	Density, γ (kN/m^2)
Concrete (EN 206)	
Lightweight Class LC 1.0	9.0 to 10.0
Normal weight	24
Cement mortar	19 to 23
Granite	27 to 30
Dense limestone	20 to 29
Softwood playwood	5.0

Table 1.
Construction material's density [5].

Categories of loaded area	qk kN/m2	Qk kN
Category A • Floors • Stairs • Balconies	1.5 to 2.0 2.0 to 4.0 2.5 to 4.0	2.0 to 3.0 2.0 to 4.0 2.0 to 3.0
Category B	2.0 to 3.0	3.0–4.0
Category C • C1 • C2 • C3 • C4 • C5	2.0 to 3.0 3.0 to 4.0 3.0 to 5.0 4.5 to 5.0 5.0 to 7.5	3.0 to 4.0 2.5 to 7.0 4.0 to 7.0 3.5 to 7.0 3.5 to 4.5

Table 2.
Imposed (live) loads [5].

- Areas for storage and industrial activities: two categories (E1 and E2)
- Garages and vehicle traffic (excluding bridges): two categories (F and G)
- Roofs: three categories (H, I, and K)

An example of variable actions (live loads) is given in the **Table 2**.

3.1.3 Wind load

The force exerted on a building by the wind is known as the 'wind load'. Wind loads are influenced by factors such as wind speed and the building's shape and surface, making them challenging to accurately predict. Structures and their elements must be analyzed and designed to withstand wind loads (EC1). These loads are determined based on the average wind speed in the region, the building's height, geometry, and surrounding obstacles that affect wind flow (**Figure 7**). Similar to

Figure 7.
Wind pressure and force direction. (a) Elevation (b) Plan (c) 3D view.

earthquake forces, wind loads are considered in two perpendicular directions and independently. It is important to note that wind and earthquake effects should not be combined in design, but rather structures should be engineered to withstand the critical impact of both loads separately. Wind pressure on a building is distributed across its exterior surface, varying based on the structure's geometry and location. Wind can cause overpressure on the windward side, potentially blowing windows in, while creating under-pressure (suction) on the leeward side, which may blow windows out. The building's shape can amplify these effects, with smooth profiles deflecting wind more effectively than textured ones, and circular buildings outperforming square shapes.

According to BS EN 1991-1-4:2005 [5], the peak pressure in Pa can be calculated using the following expression:

$$q_p(z) = C_e(z)q_b \text{ (Pa)} \tag{2}$$

$$q_b = 1/2\rho V_b^2 \text{(Pa)} \tag{3}$$

Where

$C_e(z)$ is the exposure factor (**Figure 8a**).

$q_b$is the basic velocity pressure (Pa).

ρ is the air density; $\rho = 1.25\, kg/m^{3.}$

V_b is the wind velocity, m/s (**Figure 8b**)

The wind pressure acting on the external surfaces, W_e (Pa), is obtained using

$$W_e = q_p(z)C_p(e) \tag{4}$$

where

$C_p(e)$ is the pressure coefficient for external pressure (Table 7.1 BS EN 1991-1-4:2005) and **Figure 8**.

Finally, wind force on each surface, F_w, is calculated as

$$F_w = W_eA(N) \tag{5}$$

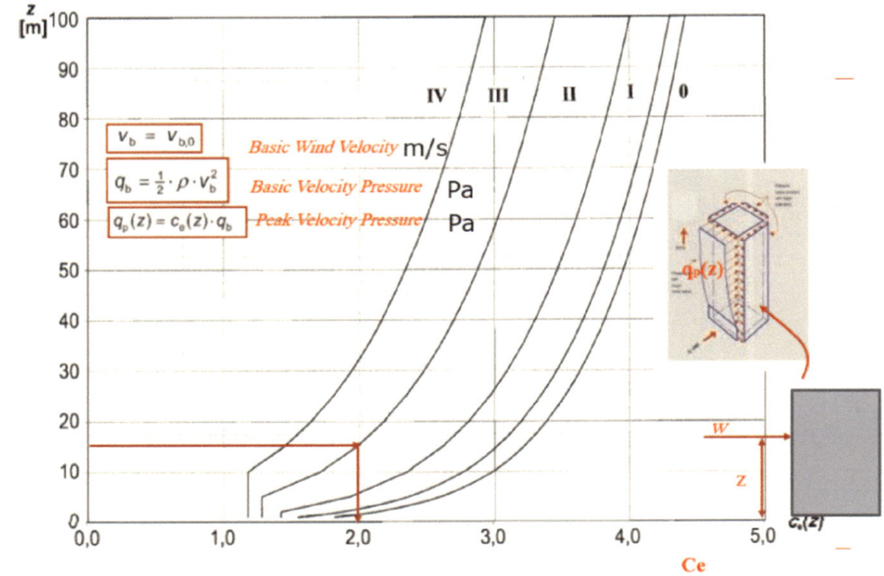

(a)

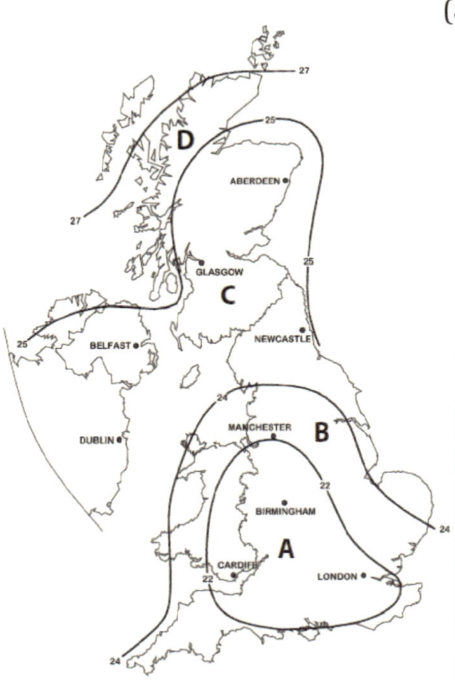

(b)

Terrain categories and terrain parameters

Terrain category	
0	Sea or coastal area exposed to the open sea
I	Lakes or flat and horizontal area with negligible vegetation and without obstacles
II	Area with low vegetation such as grass and isolated obstacles (trees, buildings) with separations of at least 20 obstacle heights
III	Area with regular cover of vegetation or buildings or with isolated obstacles with separations of maximum 20 obstacle heights (such as villages, suburban terrain, permanent forest)
IV	Area in which at least 15 % of the surface is covered with buildings and their average height exceeds 15 m
NOTE: The terrain categories are illustrated in A.1.	

(c)

Figure 8.
Wind peak pressure [5]. (a) Exposure factor Cez [5] (b) Wind velocity (c) Terrain category.

Where A is the reference area, m^2
To calculate wind load the subsequent steps must be followed:

1. Determine basic wind velocity, V_{bO} (**Figure 8b**)

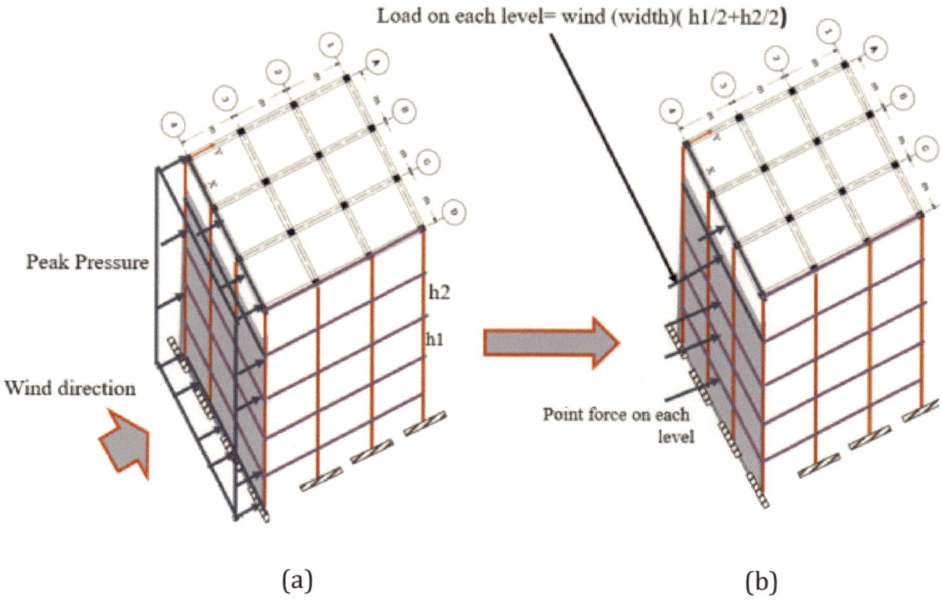

Figure 9.
Wind load on the structure. (a) Wind pressure (b) Concentrated wind force on each floor.

2. Calculate basic velocity pressure, q_b (Eq. (3))

3. Determine $C_e(z)$ (**Figure 8a**)

4. Calculate peak velocity pressure, $q_p(z)$ (Eq. (2))

5. Calculate wind load (Eq. (5) and **Figure 9**)

To analyze and design of the structures, the wind force is applied on the structure in form of point force on each floor level (**Figure 9**).

The structural configuration of the building should effectively and securely handle wind forces, transferring them to the foundations to prevent any risk of structural failure. Wind is typically identified as the primary horizontal force when assessing tall buildings through wind engineering. The structural components responsible for absorbing wind loads are typically distinct from those designed to handle dead loads and other gravity loads that originate within the building.

3.1.4 Earthquake load

Earthquakes are caused by the sudden movement of tectonic plates in the Earth's crust, which occurs along fault lines. This movement releases energy that travels through the Earth in the form of waves, causing vibrations that can be felt kilometers away from the epicenter. Areas near active fault lines are more susceptible to earthquakes.

Efforts to design structures that can withstand earthquake surface motions are continuously evolving. Alongside the guidelines provided by the EC8 Code, other codes and research teams regularly assess and update the analysis and design

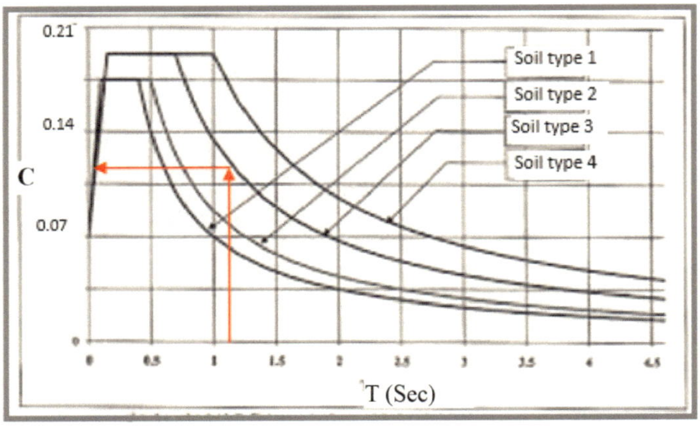

Figure 10.
Changes in the seismic response coefficient, C, with the periodicity T for various soil types.

requirements for structures against lateral earthquake loads. As a result, there are frequent changes in the building regulations regarding earthquake design.

The ACI code [6] does not specify the specific base motions for a particular site or provide detailed instructions on how the structure should be analyzed. Currently, building codes allow for the analysis of structures under the influence of various levels of earthquakes. The ASCE permits three types of structural analysis: the equivalent lateral force method, the modal spectral analysis, and the inelastic response time history analysis method.

Typically, the equivalent lateral force method is employed for buildings that are less than 50 meters in height. In some cases, it can also be used for irregular buildings, as long as all types of irregularities are carefully considered. Geotechnical studies of the soil beneath the foundation at the construction site assist designers in estimating the impact of soil type on the lateral force exerted by an earthquake.

In the equivalent lateral force method, the basic shear force can be calculated as follows:

$$V = CW \tag{6}$$

where C is the seismic response coefficient, given by code, and W is the effective seismic weight of the building. An approximate plot of the coefficient C as a function of the periodicity T is shown in **Figure 10**. For low, medium, and high-rise structures, C is approximately around 0.16, 0.06, and 0.03, respectively. For more details, refer to EC8. The approximate weight of the building can be calculated as follows:

$$W = nA(DL + \alpha LL + \Delta W_l + 3.5) + W_{walls} + a_{kh}(DL + \alpha LL) \tag{7}$$

Where

n is the number of stories

A is the floor area m^2

α is the percentage of live load participation

DL is dead load, $\mathrm{kN/m^2}$

LL is live load, $\mathrm{kN/m^2}$

ΔW_l is partitioning equivalent to overhead, kN/m^2
3.5 is the structural weight, kN/m^2
W_{walls} is the weight of all walls, kN
a_{kh} is the area of the barn, m^2

According to the author's experience, to verify the base shear force calculated by software Eq. (7) can be simplified as

$$W = 1.3nA(DL + \alpha LL + 4.5) \tag{8}$$

As the base shear force is automatically calculated by the software, Eq. (6) can provide a quick and reliable verification. For the residential buildings, the difference between software and Eq. (8) must be less than 5%.

3.2 Gravity loads distribution

The software automatically distributes the loads among the various elements. Ensuring the accuracy of this load distribution is crucial for software verification. Therefore, it is essential for all designers to have a precise understanding of gravity and lateral load distribution. After calculating the different loads, the floors must be designed before distributing the load among the frames. Once the floor/roof specifications are finalized, the gravity loads should be distributed among the frames. It is important to note that there is a fundamental difference between the distribution of gravity loads (dead and live) and lateral loads (earthquake and wind). Gravity loads, such as dead and live loads, are distributed among beams or columns based on the floor system's geometry. On the other hand, lateral loads are distributed among frames or shear walls based on their stiffness.

3.2.1 Load distribution in one-ways slabs

In one-way slabs, the load is only transferred in one direction, which means that the load on the secondary beams is directly proportional to the loading width. The calculation of each beam or joist's share will be based on its loading width. The loading width refers to the width of the slab where the loads are directly transferred to the relevant beams, which is equal to the span length between secondary beams or joist spacing. The width of the load carrier varies depending on the type of structural system of the floors and the direction of load transfer. To ensure the accuracy of the modeling and the loads inputted into the software, the load on the beams should be calculated using the presented method and compared with the values calculated by the software. The manual and software calculations should yield identical results.

The loading width of each slab, that is, a, as mentioned in the previous section, is illustrated in **Figure 11**. The load on the secondary beams of the one-way slab and the joists in the block-joist floors is a uniformly distributed load (UDL) as shown in **Figure 11** and can be determined by

$$q_D = aDL, q_L = aLL \tag{9}$$

The distribution of loads on the main beams varies between one-way and joists/ribbed systems. In the ribbed/joist system (**Figure 11b**), it is assumed that the load on the main beams is uniformly distributed.

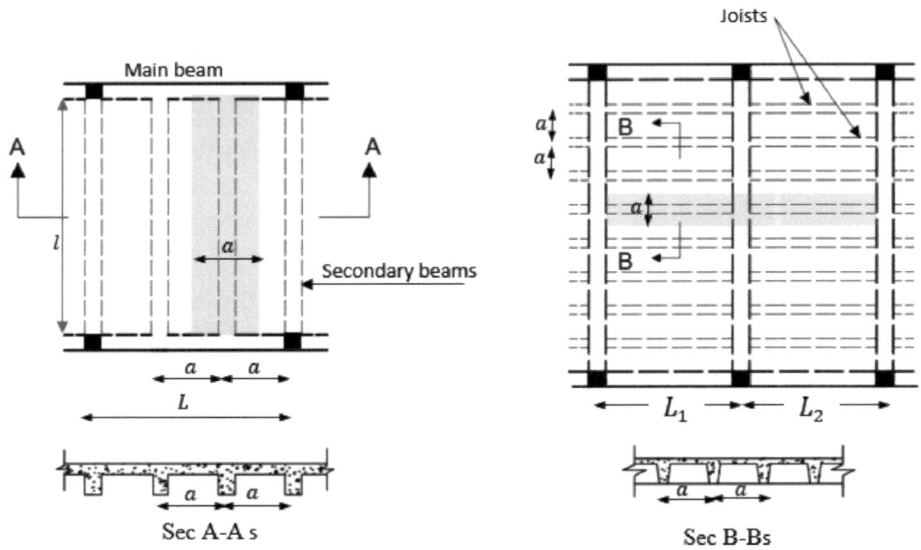

Figure 11.
Distribution of gravity loads on secondary beams and joists in one-way slab. (a) Load on the secondary beam, one-way slab; (b) Load on the joists, joist/ribbed slabs.

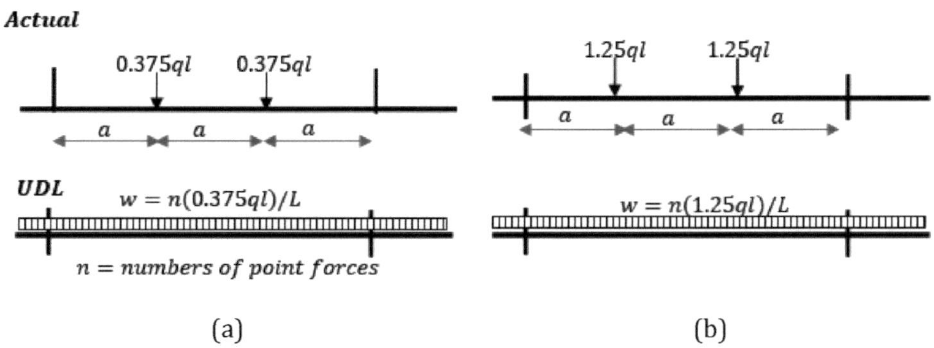

Figure 12.
Actual and approximate equivalent UDL Load on the main beam in one-way slabs (q); q is q_D or q_L (Eq. (9)). (a) Edge main beams (b) Middle main beams.

$$q_D = \left(\frac{L_1 + L_2}{2}\right)DL, q_L = \left(\frac{L_1 + L_2}{2}\right)LL \quad (10)$$

The concentrated force acting on the main beams in the one-way slabs is equivalent to the reaction supports of the secondary beams (**Figure 12**). It should be emphasized that the secondary beams are continuous, allowing for the determination of the point force on the side and middle main beams.

3.2.2 The load on the beams in two-way slabs

The load bearing surface of each beam in two-way slabs is formed by drawing the bisector of the panel corners. In the case of two-way slabs, the load is applied to the beams in a trapezoidal shape for long-span beams or triangularly for short-span beams, as depicted in **Figure 13**. These types of loads pose challenges during the approximate analysis stage, making it necessary to convert them into an equivalent uniformly distributed load (UDL). Additionally, the load caused by the weight of the walls is uniformly distributed and should be directly added to the aforementioned UDL. These calculations are crucial in validating the software modeling, which will be utilized to validate the results of the structural analysis.

The trapezoidal load on the middle long span beams, specifically axis 2 and 3, can be represented by a uniformly distributed load. This calculation can be done using **Figure 13b**.

$$q_D = \frac{(2b-a)a}{2b}DL, q_L = \frac{(2b-a)a}{2b}LL \tag{11}$$

The load distribution on the edge beams, specifically axis 1 and 4, is equal to half of the load on the middle long span beams.

Additionally, the load acting on the middle short span beams, axis B and C, is triangular in shape as illustrated in **Figure 13c** and can be calculated accordingly.

$$q_D = (a/2)DL, q_L = (a/2)LL \tag{12}$$

The load on the edge beams, axis A and D, is half of the middle beams in short span length.

3.3 Approximate structural analysis

Upon completion of the gravity and lateral load calculations, the structures undergo analysis and design to account for the impact of these loads. The analytical methods utilized include the following [7, 8]:

- Manual analysis
    - Slope-deflection method
    - Moment distribution method
    - Matrix method
- Software analysis

For each of those methods mentioned above, it is essential to have knowledge of the dimensions of the beam and column sections, while the design of these cross sections requires information on internal forces. In practical scenarios, the dimensions of the sections may need to be assumed or determined using approximate methods. The initial assumption of beam and column dimensions can result in a lengthy trial and error process. Therefore, a robust approximate structural analysis and design approach will be presented here to address this issue. These methods enable the designer to gain a comprehensive understanding of the actual internal forces and final dimensions of columns, beams, and shear walls in various project types. It is important to note that both approximate structural analysis and design techniques are

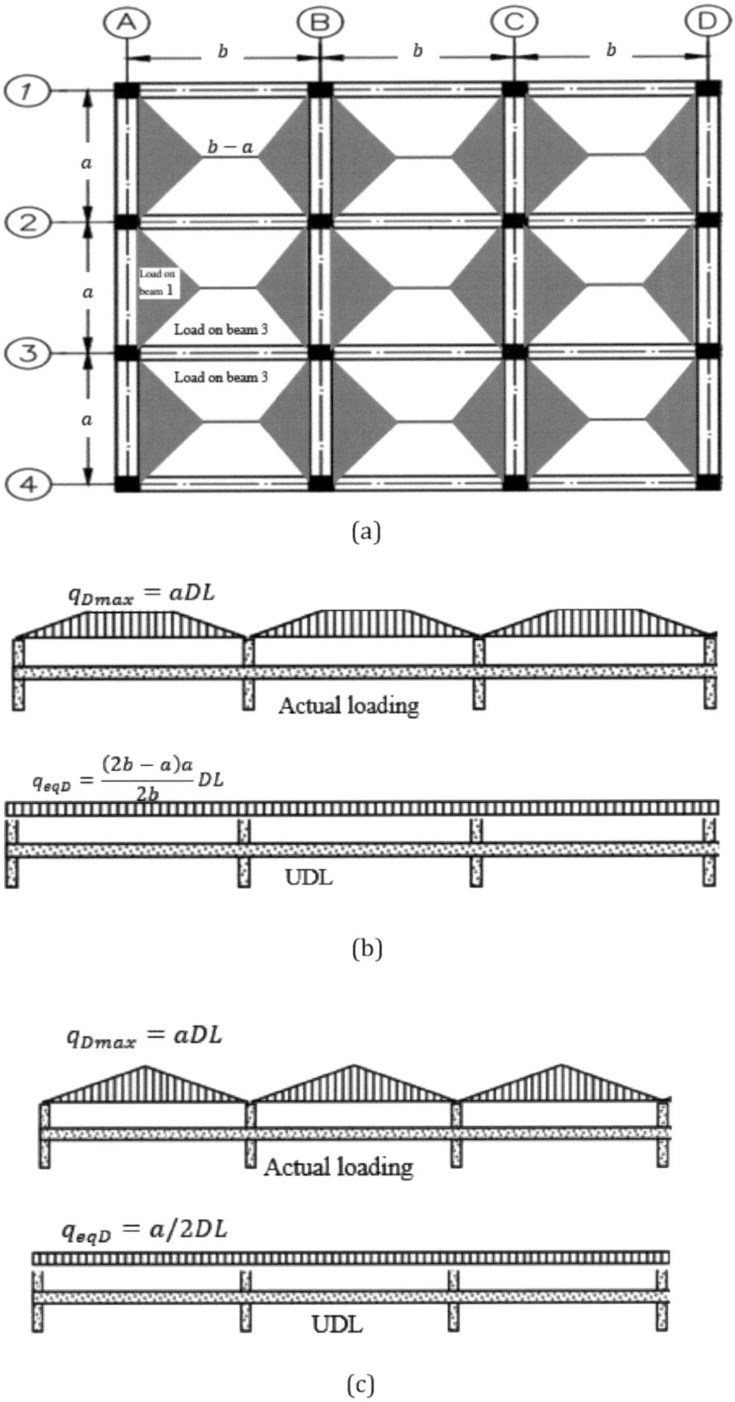

Figure 13.
Distribution of gravity loads on beams in two-way slabs. (a) Load distribution on the plan of slab. (b) Actual and equivalent UDL load on middle long span beams, that is, beam 2. (c) Actual and equivalent UDL load on middle short-span beams, that is, beam B.

straightforward, yet they serve as invaluable tools for designers in ensuring the safety of a project and the accuracy of software results.

In a pragmatic undertaking, the preliminary dimensions of the structures are initially determined through approximate analysis techniques. Subsequently, the structure undergoes a thorough analysis and design process employing precise analysis methods, either manual or computer-based. Should the sections prove unsuitable for the imposed loads, the structure is subjected to a reanalysis and redesign, incorporating new sections. This iterative process persists until the ratio of applied load to section strength falls below 1.0.

The analysis under the influence of gravity and lateral loads varies, thus necessitating a separate discussion on the specifics of each method.

It is commonly believed that only designers with a comprehensive background in structural analysis and ability to provide verifications have the authority to design structures using software. However, many analytical approaches are challenging to apply for structural verifications; therefore, the approaches presented in this chapter offer a viable alternative.

3.3.1 Approximate analysis of beam and one-way slab under the gravity loads

In the existing literature, three primary approximate analyses have been suggested for determining the initial dimensions of structural elements. These include the ACI coefficient for analyzing beams subjected to gravity loads, as well as the portal and cantilever methods for analyzing frames under lateral forces [1]. While the ACI method is recognized for its simplicity and reliability, the latter two methods tend to be more time-intensive, leading many designers to prefer alternatives.

The accuracy and reliability of the ACI coefficients method in analyzing beams and one-way slabs are evidently close to reality, making it suitable for a wide range of structures. To enhance the designer's interpretation skills and gain insight into the background of the coefficient method outlined in the ACI code [6], a continuous beam analysis with the contraflexure point assumption can provide valuable insights.

3.3.1.1 Bending moments and shear force

The position of the contraflexure point in a beam depends on its support conditions. In the both ends of a fixed beam, the contraflexure point is located at 0.2 times the length of the beam from the support. However, in a simply supported beam, the contraflexure point is located directly at the support. In a rigid reinforced concrete frame, the turning point location falls between these two values. The author proposes the position of the contraflexure point as depicted in **Figure 14**, based on the analysis of various projects. By assuming this position, the structure can be treated as a determinate system, allowing for the calculation of internal forces such as bending moments and shear forces using simple statics methods. It is important to note that the 0.1 L method, mentioned in some books, is not suitable for the approximate analysis of beams.

The bending moment analysis for the side span is shown in **Figure 15**. By applying the same method, the support bending moments at middle span length will be the same as right support of first span, that is, $\frac{wl^2}{10.8}$ and positive bending moment at the

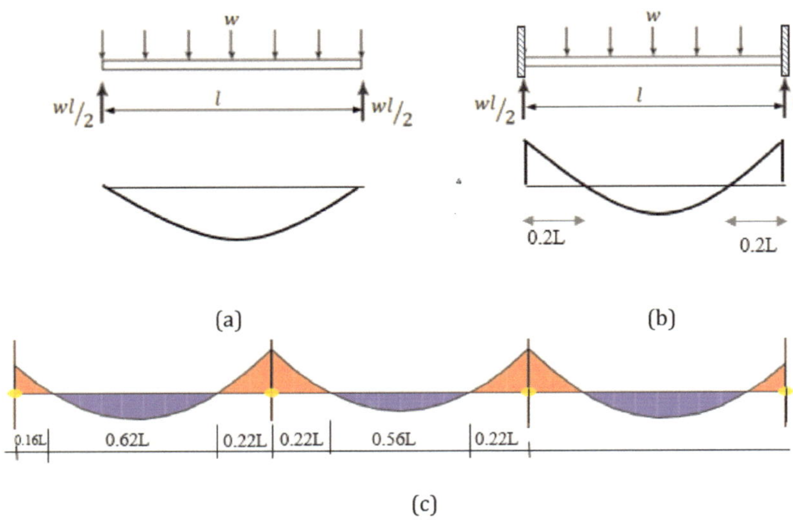

Figure 14.
Turning points in different span lengths, suggested by the author. (a) Simply supported beams. (b) Fixed beams. (c) Continuous beams.

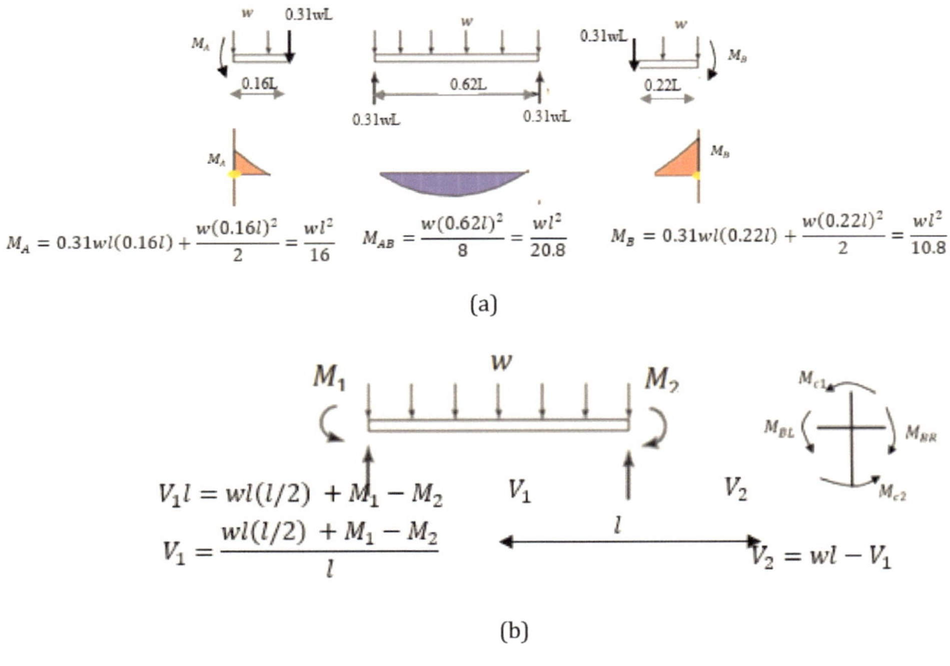

Figure 15.
Bending moment and shear force analysis in the first span length. (a) Bending moment. (b) Shear force.

middle of the beam will be $\frac{wl^2}{20.8}$. In this method, the bending moment in the columns is calculated from the equilibrium of moments at the joints

$$M_{c1} = M_{c2} = (M_{BL} - M_{BR})/2 \tag{13}$$

Eq. (13) demonstrates that when subjected to gravity load, the bending moments in the central columns with identical beam span lengths will be null. However, for beams with different span lengths, the bending moments in the middle columns will still be minimal. It is important to note that this scenario only holds true when the live load is uniformly distributed across all span lengths. In all cases, the bending moment in the side columns remains significant.

In order to obtain a more thorough and accurate estimation, the influence line method must be utilized to compute the critical bending moments at different positions under live load. The ACI coefficient method [6] has employed the influence line to determine these critical bending moments. For negative bending moments, it is assumed that the live load is distributed across both adjacent span lengths. Conversely, for positive bending moments, the live load is assumed to be applied solely on the span where the bending moment is to be determined, with the other spans being loaded in between [9].

The ACI code [6] offers a set of coefficients that can be utilized to analyze continuous beams and one-way slabs subjected to gravity loads. In the event that certain conditions are satisfied, the method of statutory coefficients can be employed.

- The number of span lengths are equal to 2 or more
- The length of the spans should be almost equal, or in two adjacent spans, the length of the longer span does not exceed 20% of the length of the shorter span length
- Loads are uniformly distributed
- The live load should not be greater than three times the dead load; $LL \leq 3DL$
- The beam's moment of inertia should be constant

The limitations imposed on span lengths and maximum live load are in place to guarantee that the positioning of live load (in order to optimize the bending moment within a specific section) on the spans will not result in negative bending moments in sections typically experiencing positive bending moments (refer to **Figure 16**) [10].

ACI code allows the calculation of bending moment and shear force in continuous beams and one-way slabs using formulas that obtain the internal forces at critical points

$$M = \alpha w l_n^2 \tag{14}$$

Where

M is bending moment, kNm

α is the ACI coefficients, **Figure 17**

w is uniformly distributed dead or live load on the beam, kN/m

l_n is the clear span length, m **Figure 18**

The coefficients for continuous beam and one-way slab are the same, except for negative bending moment at the side supports (**Figure 17b**).

The comparison between the approximate analysis suggested by the author (**Figure 15**) and the ACI method reveals that the positive bending moments derived from the ACI coefficients method tend to be somewhat conservative due to the consideration of the influence line.

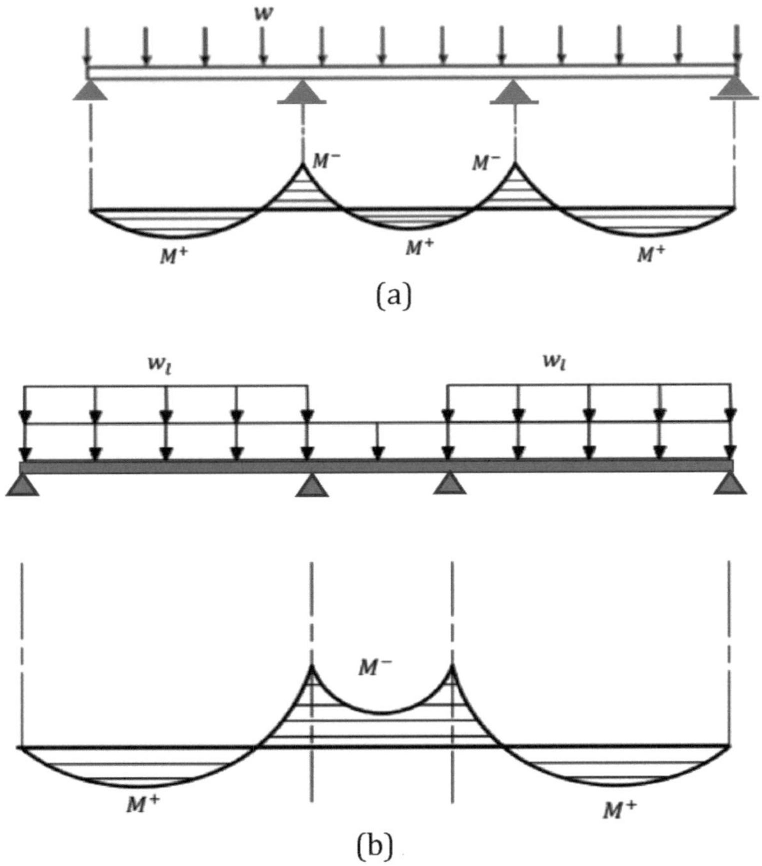

Figure 16.
The effect of span length on the shape of the bending moment diagram. (a) Approximate analysis using ACI coefficients is allowed. The UDL load applied to the span is almost identical, resulting in a positive bending moment in the middle of the spans. (b) Analysis using ACI coefficient method is not allowed, unequal spans and type of live loading caused negative bending moment throughout the middle span.

For slabs with a span of less than 3 meters and for beams where the ratio of stiffness of the column to the stiffness of the beam – at both ends – exceeds 8, the negative bending moment at the edge of the supports can be approximated as $w_u {l_n}^2/12$ instead of the values provided in **Figure 17**. By utilizing the bending moments at the ends of each span, the shear force can be determined using the approach outlined in **Figure 15**. Additionally, a conservative estimate of the shear force is presented by ACI

$$V = \beta w l_n \tag{15}$$

where V is shear force and β is ACI coefficient for shear force, **Figure 18**.

When utilizing code coefficients for beam analysis, it is necessary to calculate the bending moment of the columns, M_{CT} *and* M_{CB}, individually **Table 3**. Assuming that the larger span length next to the column sustains half of the live load in addition to the dead load, while the smaller span length only carries the dead load, the bending moment transmitted to the columns can be determined using the provided formula

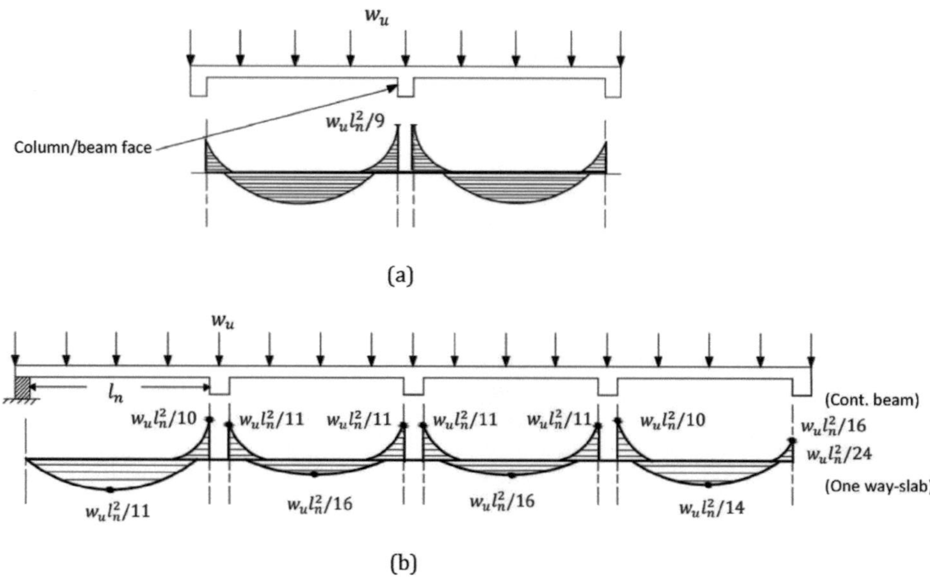

Figure 17.
Approximate analysis under gravity loads – continuous beam/one way slab – ACI coefficient for bending moments, α [2]. (a) Bending moment – two spans length (b) Bending moment – continuous beam or one-way slab.

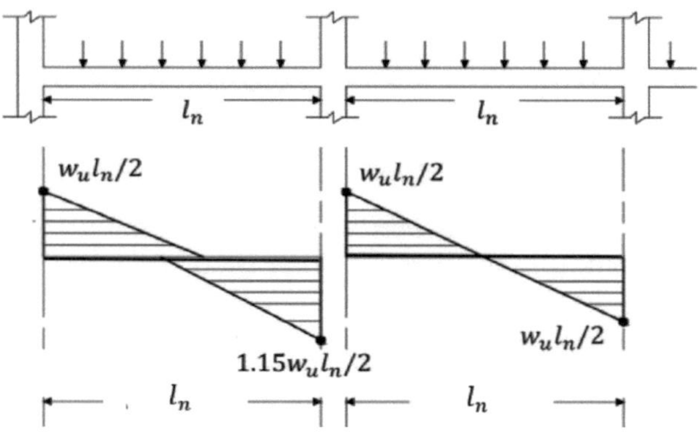

Figure 18.
Shear force values in the face of supports.

L_{n2}	Bigger span length
L_{n1}	Smaller span length
w_D	Uniformly distributed dead load
w_L	Uniformly distributed live load

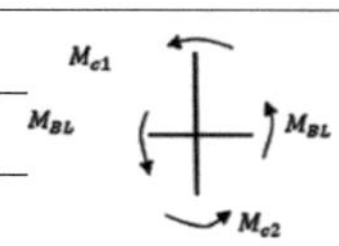

Table 3.
Beam span length definitions.

$$M_{CT} = M_{CB} = 0.035\left[(w_D + 0.5w_L)L_{n2}^2 - w_D L_{n1}^2\right] \tag{16}$$

It should be emphasized that, in the majority of structures, the bending moment in middle columns under gravity load is relatively minor in comparison to the bending moment caused by lateral forces. Conversely, the axial force in columns under gravity loads is significant when compared to the axial force resulting from lateral forces.

3.3.1.2 Approximate axial force in the columns

The calculation of the axial force for each column involves multiplying the effective loading area of the column by the applied loads. The effective loading area of the column on each floor is determined by considering the center line of the panels surrounding the columns, as shown in **Figure 19**. To determine the axial force of each column on a specific story, the axial forces of the columns on the upper stories are summed up until the desired story is reached. This approach is suitable for frames with simple connections, but it may result in a significant approximation for rigid frames. To obtain more realistic axial forces for rigid frames, according to analytical approaches and the results of software analyses, the author proposes an approximate method for calculating the axial force of the columns:

$$P_D = 0.4n\left(f+\frac{a}{2}\right)\left(j+\frac{L_1}{2}\right)DL + P_{wall} + P_{Col} P_L = 0.4n\left(f+\frac{a}{2}\right)\left(j+\frac{L_1}{2}\right)LL \text{Corner columns, A1} \tag{17}$$

$$P_D = 1.25n\left(\frac{b+c}{2}\right)\left(\frac{L_1}{2}\right)DL + P_{wall} + P_{Col} P_L = 1.25n\left(\frac{b+c}{2}\right)\left(\frac{L_1}{2}\right)LL \text{Edge columns, C1} \tag{18}$$

$$P_D = 1.25n\left(\frac{a+b}{2}\right)\left(\frac{L_1+L_2}{2}\right)DL + P_{wall} + P_{Col}\, P_L = 1.25n\left(\frac{a+b}{2}\right)\left(\frac{L_1+L_2}{2}\right)LL \text{Edge columns, B2} \tag{19}$$

where n is the number of floors above the column in the target story. The weight of the walls is also obtained from the product of the UDL load of the wall by half the length of the span lengths adjacent to the column (P_{wall}). It should be noted that the

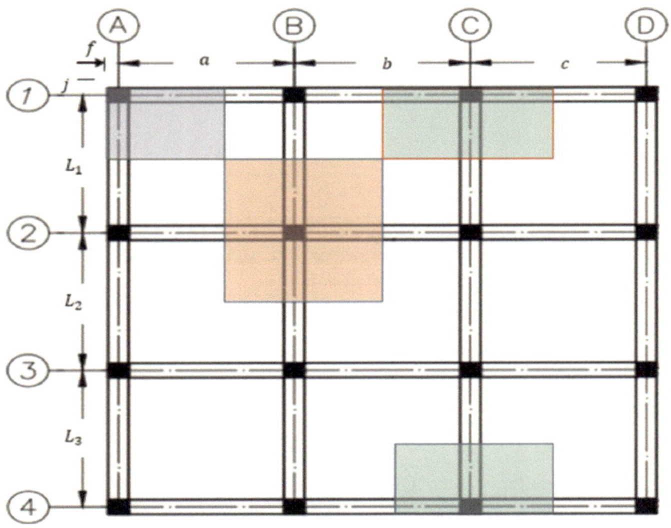

Figure 19.
Effective loading area of columns.

weight of the column itself should also be considered; product of column dimensions*concrete density*height of the column*numbers of columns

3.3.2 Approximate analysis of frames under lateral loads

Two methods are employed to examine the response of frames to the horizontal forces induced by wind or seismic activity [11]

- Portal method
- Cantilever method

Explanations regarding the methods and their advantages and disadvantages can be found in structural analysis books. However, this discussion will focus solely on the portal method, which is considered to be simpler and more practical. The principle of this method is based on assuming that the contraflexure point is located at the midpoint of the beam and column. It should be noted that this assumption leads to an underestimation of the bending moment on the ground floor by at least 30%, although the results for other floors are deemed acceptable. Despite its practicality, this method is not commonly used in practical analysis and design due to its time-consuming nature.

To comprehend the fundamentals of the portal method, a single-span frame is employed. By separating the beam and column at the turning point, the indeterminate structure is transformed into a determinate structure. The turning point is assumed to be situated in the middle of both the column and beam. It is worth mentioning that this assumption has been proven to be accurate for all floors except the ground floor (**Figure 20**).

The distribution of shear force in each floor is determined by the stiffness ratio of the columns in each story. Initially, the dimensions of the columns are unknown, so the shear force is divided based on the loading span. As a result, each column experiences a shear force of P/2. Referring to **Figure 20c**, the bending moment of the column can be expressed as $M_c = P/2(h/2)$. At point B, the sum of beam bending moments is equal to the sum of column bending moments, denoted as $\sum M_{Bi} = \sum M_{ci}$, so $M_B = M_c$ Consequently, M_B is equal to M_c. By applying equilibrium in the beam, as shown in **Figure 20c**, we can derive the necessary equations:

$$\frac{V_B L}{2} = M_B \tag{20}$$

Consequently, the beams' shear force will be equivalent to the beam bending moment divided by half of the beam span length

$$V_B = \frac{M_B}{L/2} \tag{21}$$

Based on the equilibrium of forces in the vertical direction at point B, it can be determined that the axial force of the column is equal to the shear force of the beam, denoted as $N = V_B$. Consequently, the step-by-step procedure for analyzing the frame using the portal method is outlined as follows:

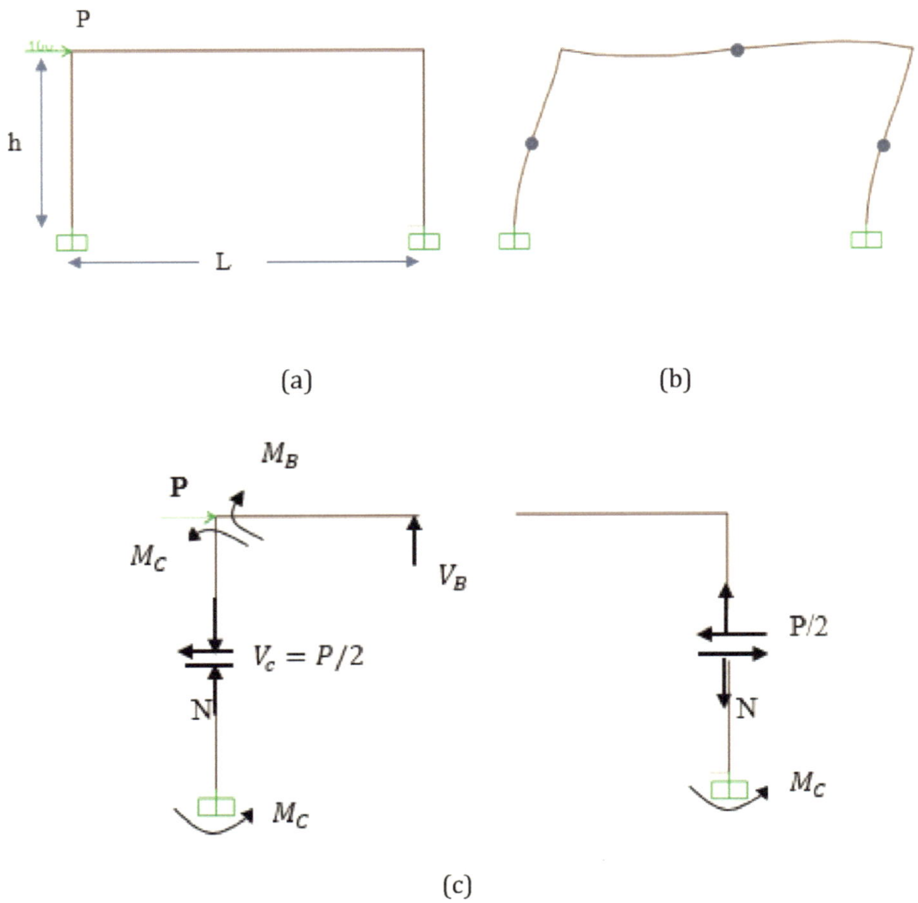

Figure 20.
The principles of the portal method. (a) One-span frame. (b) Turning points in the flexural frame. (c) Internal forces.

1. The determination of the turning points of the beam and column involves establishing their respective locations. Under an acceptable hypothesis, it is observed that the turning point on all floors, except for the ground floor, is situated in the middle of the beams and columns.

2. The shear force in each floor is computed by summing up the lateral force exerted on the floors from the roof up to the desired story.

3. The shear force of each column is equal to the loading span of the column divided by the length of the frame, multiplied by the shear force in the corresponding story.

4. The bending moment of each column is determined by multiplying the shear force of the column by half the height of the column.

5. The bending moment of each beam is calculated by considering the equilibrium of joints

6. The calculation of the shear force for each beam involves dividing the end bending moment by half of the span length.

7. The axial force in each column is determined by adding the axial force of the columns in the upper floors to the variation in shear force of the adjacent beams.

8. In this approach, the shear force in the beams is uniform, resulting in only the side columns experiencing axial force while the middle columns remain unaffected by the earthquake.

$$M_{c1} + M_{c2} = M_{BL} + M_{BR}$$

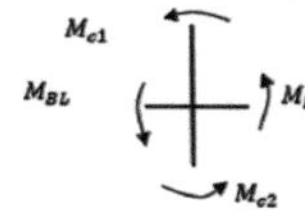

It is to be noted that, author suggest that, the axial force of other columns can be calculated using the similarity of triangles:

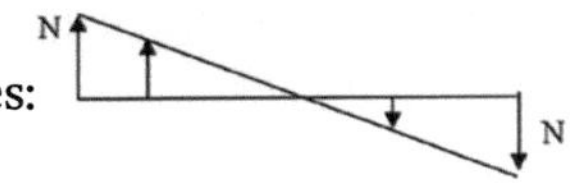

3.3.2.1 New method suggested by the author

The portal method is characterized by its length and the numerous calculations it entails. In contrast, the author suggests a straightforward approach akin to the ACI coefficient method for determining shear force and bending moment in columns subjected to lateral loads from earthquakes. Typically, columns within each story exhibit similar dimensions in practical applications. With the assumption of a rigid floor, the stiffness of all ground floor columns will be uniform, leading to nearly equal shear forces in the columns.

$$V_c = \frac{V}{n} \tag{22}$$

In the given scenario, V represents the story shear force, V_c denotes the shear force in each column, and n signifies the number of columns in the story. On the ground floor, the turning point is positioned at a distance of 0.8 h from the support. Consequently, the bending moment in the ground floor in side and middle columns is equal to

Side columns

$$M_c = 0.8hV_c \text{ in support} \tag{23}$$

$$M_c = 0.2hV_c \text{ At the top of the column} \tag{24}$$

middle columns

$$M_c = 0.7hV_c \text{ in support} \tag{25}$$

$$M_c = 0.3hV_c \text{ At the top of the column} \tag{26}$$

In alternative narratives, as per software analyses conducted on several buildings with 5–15 stories, it is evident that the shear force exerted on the side and middle columns is not equivalent

$$V_c = 0.73\frac{V}{n} \text{ Side columns} \tag{27}$$

$$V_c = 1.19\frac{V}{n} \text{ Middle columns} \tag{28}$$

Since in other stories the turning point is in the middle of the column, the bending moment in the columns is equal to

$$M_c = \frac{V_c h}{2} \tag{29}$$

The simplified method mentioned above is employed to determine the bending moments of the upper and lower columns under the influence of earthquake or wind load. Additionally, the bending moments on the beams can be computed by utilizing joint equilibrium. Typically, calculations commence from the left side, resulting in the bending moments of the beam being equal to the bending moments in the side joints

$$M_{BR} = M_{c1} + M_{c2} \tag{30}$$

In the rest of the middle joints, the beam bending moments is equal to

$$M_{BR} = M_{BL} = (M_{c1} + M_{c2})/2 \tag{31}$$

The calculation of the shear force in the beams involves dividing the bending moment of the beam by half of its length (Eq. (21)). It is crucial to note that the calculations provided consider the scenario where the slab does not contribute to resisting lateral loads (membrane floor).

The simplified method mentioned above can be utilized as a suitable alternative to the portal method for approximate analysis during the initial design phase and verification of dimensions when inputting data into software tools.

It is important to highlight that the aforementioned method has been introduced under the premise of a symmetrical structural configuration, characterized by uniform column and beam dimensions across each level. While this assumption introduces a degree of approximation in the results, it remains a valuable approach for establishing the preliminary dimensions of structural components. Furthermore, it serves as a robust tool for validating the structural modeling, as well as the analysis and design processes performed by software.

3.4 Preliminary design of structural elements

The software utilizes two methods to select the initial dimensions as input data. The first method involves assuming initial dimensions and optimizing them through trial and error using the software. The second method is based on approximate analysis, while also considering the code limit on the beam depth.

The process of performing approximate analysis and designing is crucial. By conducting a preliminary design based on approximate analysis, the designer gains valuable insights into the final dimensions of the beams, columns, shear walls, and foundations. Additionally, approximate analyses assist the design engineer in interpreting the results of computer analyses and ultimately validating them.

3.4.1 Beams

The ACI coefficients method is employed to calculate the bending moment on the beams at various points (typically on the first middle column) under the dead and live

	h_{min}			
	Simple support	**One end cont.**	**Both end cont.**	**Cantilever**
One-way slab	$l/20$	$l/24$	$l/28$	$l/10$
Beam	$l/12$	$l/15$	$l/15$	$l/6$

Table 4.
Minimum beam depth [6].

load, for the preliminary design of the dimensions (b, h). On the other hand, the author's simplified method is utilized to determine the bending moment in the columns.

By considering the loading combination of 1.2D + 1.0 L + 1.0E, the factored bending moment of the beam is computed at the first middle support (**Table 4**). Consequently, the minimum cross section dimensions of a beam with a singly reinforced section are determined to be

$$bd^2 = \frac{M_u}{0.2f_c}, b = 0.65d, h = d + 55mm \tag{32}$$

As stated in chapter 8 [2], the regulatory approach offers a more straightforward solution for determining the initial dimensions (**Table 3**). The constraints presented in **Table 3** are primarily intended to regulate the deflection caused by gravitational loads. However, given the significant influence of earthquake forces on beam design, it is recommended by the author to establish the fundamental dimensions in order to effectively manage lateral deformation and accommodate sufficient space for reinforcement bars

Braced frame

$$\text{Depth} \geq \text{h}_{\text{min}}\ b = 0.65h \text{ Width} \tag{33}$$

Unbraced frame

$$\text{Depth} \geq 1.5\text{h}_{\text{min}}\ b = 0.65h \text{ width} \tag{34}$$

The final dimension would be the maximum of those two approaches.

3.4.2 Columns

The bending moment caused by dead and live loads is considered insignificant, with the exception of the side columns. For the middle columns, the axial force resulting from dead and live loads, as well as the bending moment in the columns due to earthquake, are calculated using the author's proposed method of 1.2D + 1.0 L + 1.0E.

In practical design, it is advisable to incorporate earthquake/wind force when calculating the initial dimensions of columns. To achieve this, the author proposes a straightforward approach outlined by author [2], which involves utilizing the code limitations on lateral displacement. According to chapter 8 [2], the minimum moment of inertia of each column is as follows

Unbraced frame

$$I > 12\frac{VC_d h_s^2}{nE} \text{ Less than five floors} \tag{35}$$

$$I > 14.5\frac{VC_d h_s^2}{nE} \text{ More than five floors} \tag{36}$$

Braced frames

Given a similar approach and assuming that the frames can withstand a maximum of 30% lateral load, the calculation for determining the minimum moment of inertia of each column is as follows:

$$I > 3.6\frac{VC_d h_s^2}{nE} \text{ Less than five floors} \tag{37}$$

$$I > 4.5\frac{VC_d h_s^2}{nE} \text{ More than five floors} \tag{38}$$

where

$$I = \frac{bh^3}{12}$$

V= shear force in the story
C_d= displacement magnification factor, for example, $C_d = 5.5$
h_s= story height
E= elastic module
n= numbers of column in the story
b=column dimension parallel to the bending moment axis
h= column dimension normal to the bending moment axis

The effectiveness of the proposed method has been confirmed through a range of real-world projects involving 5- to 15-story reinforced concrete structures over the past two decades, all of which were designed by the author.

The dimensions of b and h are carefully selected by the designer in order to provide the required moment of inertia, $I = \frac{bh^3}{12}$. To simplify matters, it is advisable to initially consider the columns as being square in shape. In the initial stages of a project, the columns on each story can be treated as identical. However, during the optimization stage, their dimensions can be adjusted according to the internal forces using specialized software.

The following simplified method can also be used to calculate the initial dimensions of the columns

Column under axial force and bending moment

$$bh \geq \frac{P_u}{0.3f_c} \tag{39}$$

Column under axial force only

$$bh \geq \frac{P_u}{0.7f_c} \tag{40}$$

3.5 Case study 1: Structural analysis and design; National Housing: Kurdistan - Bahran Sanandaj 2020–2022

The National Housing project represents a residential development situated in the outskirts of Sanandaj, within the Kurdistan province. This project comprises 35 blocks, categorized into 15 distinct types, with each block containing between 5 to 10 units. A standard architectural layout of the floors is illustrated in **Figure 21**.

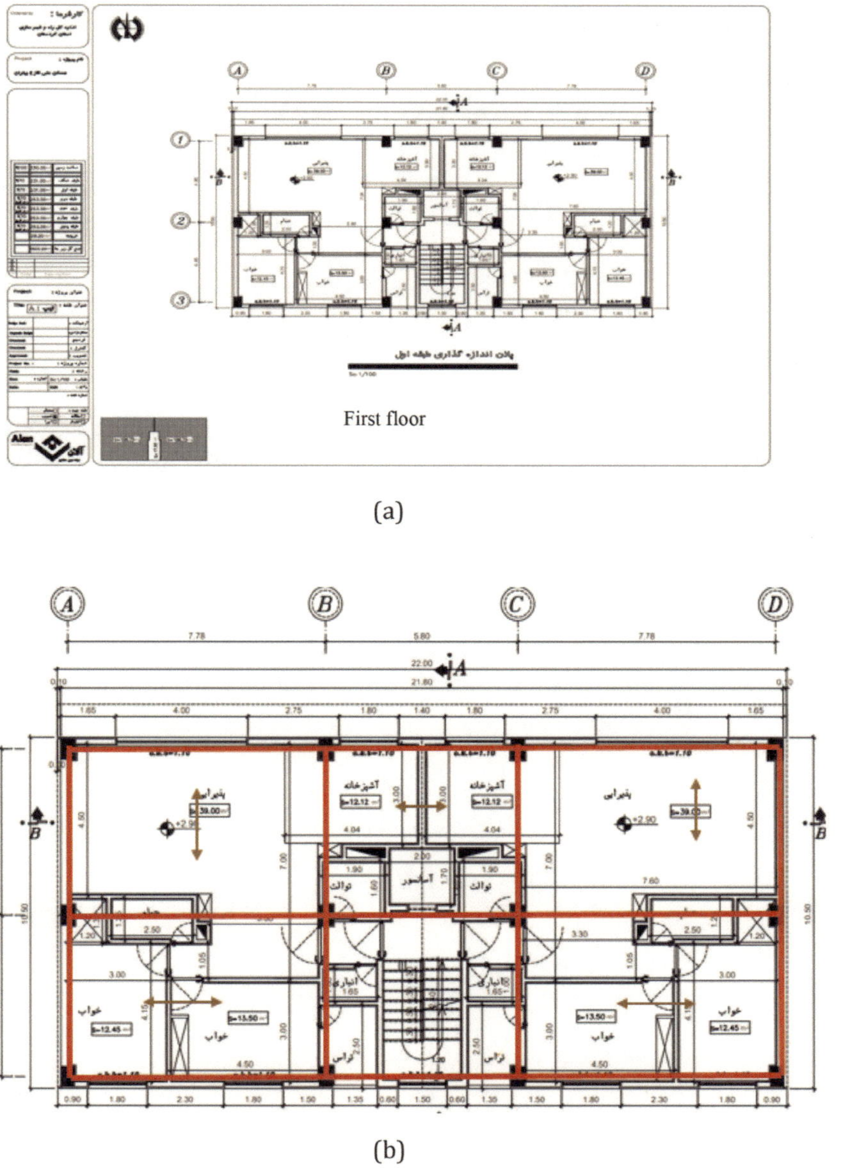

Figure 21.
The National Housing project. (a) Typical floor plan – first floor (b) Column/beam layout, and floor joist direction.

3.5.1 Summary of loading

<table>
<tr><td rowspan="5">(KN/m^2)</td><td>Location</td><td>Dead</td><td>Live</td><td>Partition</td><td colspan="2">Snow</td></tr>
<tr><td rowspan="2">Roof</td><td rowspan="2">4.99</td><td rowspan="2">1.5</td><td rowspan="2">-</td><td>Distributed</td><td>1.26</td></tr>
<tr><td></td><td></td></tr>
<tr><td>Story2-6</td><td>5.01</td><td>2.0</td><td>1.0</td><td colspan="2">-</td></tr>
<tr><td>balcony</td><td>5.01</td><td>3.0</td><td>-</td><td colspan="2">-</td></tr>
</table>

<table>
<tr><td rowspan="6">Wall Load</td><td>Location</td><td>height</td><td colspan="2">Wall Type</td><td>KN/m^2</td><td>KN /m</td></tr>
<tr><td rowspan="2">Roof</td><td rowspan="2">0.85</td><td colspan="2">External with Cladding</td><td>-</td><td>3.15</td></tr>
<tr><td colspan="2">External without Cladding</td><td>-</td><td>2.7</td></tr>
<tr><td rowspan="3">Story</td><td rowspan="3">2.9</td><td rowspan="2">External with Cladding</td><td>With Opening</td><td>2.31</td><td>7.6</td></tr>
<tr><td>Without Opening</td><td>3.3</td><td>10.89</td></tr>
<tr><td colspan="2">External without Cladding</td><td>2.8</td><td>7.8</td></tr>
</table>

3.5.2 Approximate structural analysis

Performing approximate analyses and creating designs according to these findings is crucial. Approximate analysis aids the designer in understanding the outcomes of the computer analysis and ultimately confirming their accuracy. The approximate analyses for both gravity and lateral loads are conducted under various scenarios.

3.5.2.1 Gravity load

As the span lengths of beams are approximately the same, ACI method provides acceptable results. In the beam 2 (**Figure 21b**), maximum bending moment at the first middle column assuming

$$M_{DL} = \frac{w_D l_n^2}{9} = \left(\frac{20(4.75)^2}{9}\right) = 50.13\ kNm$$

$$M_{LL} = \frac{w_L l_n^2}{9} = \left(\frac{8.0(4.75)^2}{9}\right) = 20.06\ kNm$$

where

$$w_D = 8.0/2[(5.0)] = 20\ kN/m$$

$$w_L = 8/2[(2.0)] = 8.0\ kN/m \text{ and } l_u = 4.75$$

3.5.2.2 Lateral forces

The seismic load calculation involves categorizing the country of Iran into four distinct zones. The specific project is situated in Sanandaj, Kurdistan, an area characterized by a significant earthquake risk, with a seismic coefficient of A = 0.35 g. An RC frame exhibiting intermediate ductility has been selected for this project. The base shear force is determined accordingly.

$$V = CW \geq 0.12\mathrm{AIW}$$

$$C = \frac{ABI}{R} = 0.157$$

$$W = 1.3nA(DL + \alpha LL + 4.5)$$

$$W = 17{,}335.89\ kN$$

$$V = 0.157\ (15{,}554) = 2{,}722\ kN$$

For simplicity, to analyze structure under lateral loads, the method proposed by authors is used.

According to the proposed method and considering 12 columns in each floor (**Figure 21**), shear force in each ground column is

$$V_{c,s} = 0.9\frac{V}{n} = 0.9\ \frac{2722}{12} = 204.15\ kN\ Edge\ column$$

$$V_{c,m} = 1.20\frac{V}{n} = 1.20\ \frac{2592.3}{12} = 272.20\ kN\ Middle\ columns$$

Assuming the contraflexure point at 0.8 h from the support, the bending moment in the ground floor is equal to

Edge column:

$$M_{c,bot} = 0.8hV_c = 0.8(3.1)204.15 = 506.29\ kNm\ at\ the\ support$$

$$M_{c,top} = 0.2hV_c = 0.2\ (3.1)204.15 = 126.57\ kNm\ at\ the\ top\ of\ column$$

Middle column:

$$M_{c,bot} = 0.7hV_c = 0.7(3.1)272.20 = 590.67\ kNm\ at\ the\ support$$

$$M_{c,top} = 0.3hV_c = 0.3\ (3.1)272.20 = 253.14\ kNm\ at\ the\ top\ of\ column$$

In the first floor ($V = 2592.3\ kN$), also the shear forces in the side and middle columns are not equal

$$V_{c,s} = 0.70\frac{V}{n} = 0.70\ \frac{2592.3}{12} = 151.21\ kN\ side\ column$$

$$V_{c,m} = 1.2\frac{V}{n} = 1.2\ \frac{2592.3}{12} = 259.23\ kN\ middle\ columns$$

Since in other floors, the contraflexure point is in the middle of the column, the bending moment at the first floor and in the side columns are equal to

$$M_c = \frac{V_c h}{2} = \frac{151.21(3.0)}{2} = 226.81\ kNm$$

Using equilibrium in Joint A2, the bending moment of the beam is

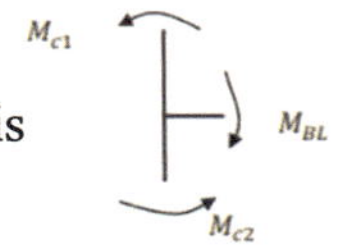

$$M_{BL} = M_{c1} + M_{c2} = 126.57 + 226.81 = 353.38\ kNm$$

Bending moment in the middle column of the first floor is

$$M_c = \frac{V_c h}{2} = \frac{259.23(3.0)}{2} = 388.85\ kNm$$

The bending moment of the beams in the left and right hand of joint B2 is equal to

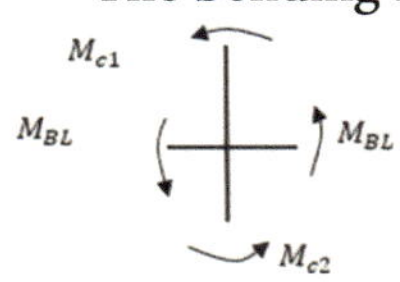

$$M_{BR} = M_{BL} = \frac{(M_{c1} + M_{c2})}{2} = \frac{253.14 + 388.85}{2} = 321.00\ kNm$$

The summary of the results is shown in **Figure 22**.

3.5.3 Preliminary design of structural elements

3.5.3.1 Beams

According to Eq. (32), the minimum dimensions of the beam B (**Figure 22**) assuming singly reinforced section are equal to

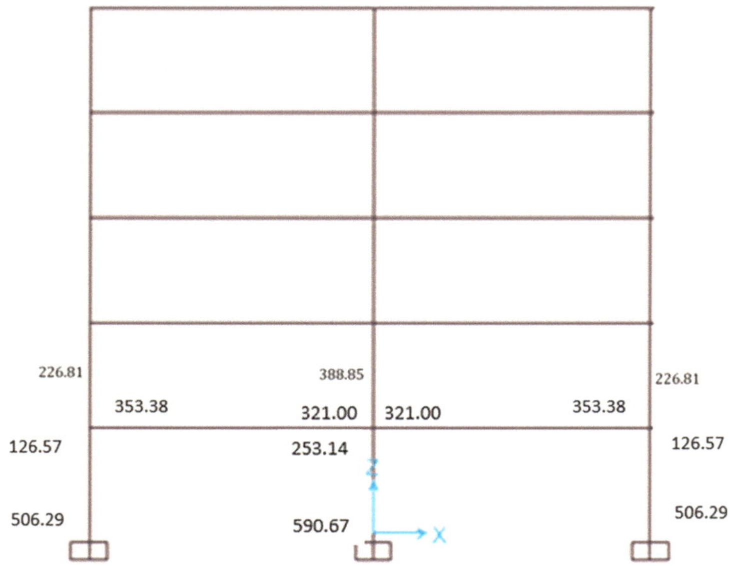

Figure 22.
Proposed method; approximate analysis – bending moment under lateral force kNm – Fame B.

	Story	G, 1	2,3	4,5
Columns	550(550)	450(450)	400(400)	mm
Beam	350(550)	350(450)	300(400)	mm

Table 5.
Summary of initial design.

$$bd^2 = \frac{M_{Ed}}{0.2f_c}, b = 0.65d, h = d + 55mm$$

$$Load\ combo : 1.2D + L + E \text{ (Iranian code)}$$

$$M_{Ed} = 1.2(50.13) + 20.06 + 353.38 = 433.60kNm$$

$$bd^2 = \frac{433.60E6}{0.2(30)}, b = 0.65d, d = 480.55\, mm$$

$$b = 0.65d = 0.65(480.55) = 312.35\, mm$$

$$h = 480.55 + 55 = 535.55\, mm,$$

$$use\ h = 550\ mm$$

$$b = 350\ mm$$

Furthermore, according to **Table 3**, minimum depth of the beam in unbraced fame system in seismic zone is L/18.5 for continuous beam (ACI/Iranian code). To consider the effect of seismic for on the dimensions author propose Eq. (33):

$$\text{Effective deptd} \geq 1.5\text{d}_{\text{min}} = 1.5\left(\frac{4.55}{18.5}\right) = 368.91\, mm$$

$$h = 368.91 + 55 = 423.91\, mm, h = 450\ mm$$

$$b = 0.65h = 0.65(450) = 292.5\, mm, b = 300\ mm$$

Final dimension based on design: 350(550) mm

3.5.3.2 Columns

In a practical design, it is better to calculate the initial dimensions of the columns considering the earthquake force **Table 5**. For this purpose, the author suggests a simple method.

In an Unbraced six story RC frame, the initial dimension can be calculated as follow

$$I > 14.5\frac{VC_d h_s^2}{nE} = 14.5\frac{2722(1000)4.5(3200)^2}{12(25000)} = 60624384400\ mm^4$$

$$\frac{bh^3}{12} = 60624384400, b = h = 519.35\, mm\ use\ b = h = 550\ mm$$

3.5.4 Verifications

Manual analysis skills are considered essential for a designer. In the absence of dependable manual analysis to validate software analyses and modeling, software

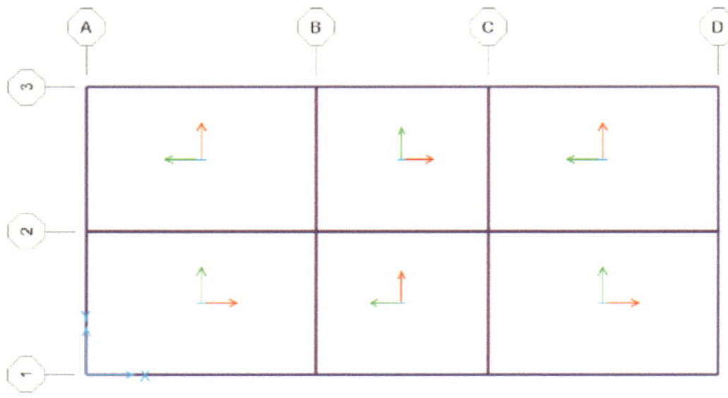

(a)

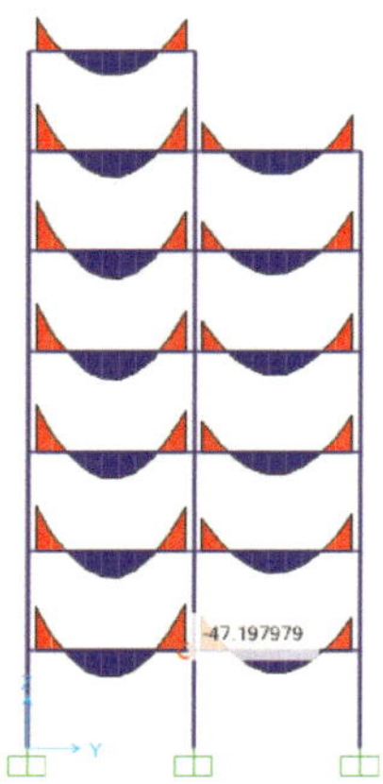

(b)

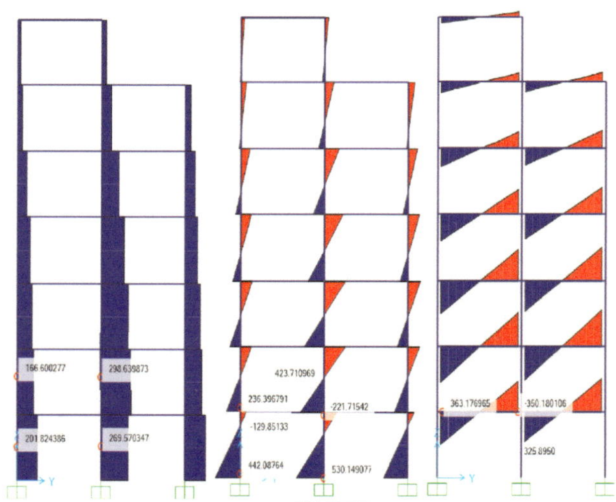

(a) (b) (c)

Figure 23.
Software analysis under seismic load. (a) Plan and joists direction (red arrows). (b) Bending moment under dead load (permeant action): Frame B. (c) Proposed method: $M_{DL} = 50.13$ kNm Shear force in columns (Kn) (d) Bending moment in columns, kNm (e) Bending moment in beams, kNm.

analysis and design may result in a catastrophe. In order to validate the loading, structural modeling, and reliability of the analyses, the software analysis (**Figure 23**) was compared with approximate analyses proposed by the author (**Figure 22**). The results show that the difference is between 1 and 12%. Thus, it can be inferred that the modeling and results obtained by software are deemed acceptable.

The comparison of the bending moments under gravity load derived from the proposed method, which yields $M_{DL} = 50.13\ kNm$, and the software analysis, which results in $M_{DL} = 47.19\ kNm$ (as illustrated in **Figure 23a**), indicates a discrepancy of 5.8%.

Additionally, a comparison between the approximate analysis (**Figure 22**) and the software analysis (**Figure 23**) conducted under lateral seismic load reveals discrepancy in columns ranging from (129.85–126.57)/129.85 = 2.5% to (590.67–530.14)/590.14 = 10.03%. Additionally, for the beams, the variation is observed to be between 2.7 and 8.22%.

3.6 Approximate analysis and design of shear wall

When selecting the dimensions of the shear wall section, it is imperative to ensure that it meets the necessary criteria:

a. The section must possess adequate strength to withstand the applied factored axial force, bending moment, and shearing force.

b. The section should also exhibit adequate lateral stiffness to restrict lateral displacement within the acceptable limits stipulated by the relevant building code.

There is no universally accepted approach for selecting the dimensions of a shear wall (**Figure 24**). To establish a suitable criterion for the initial design of these dimensions, the author suggests an approximate method [[2], chapter 11]. This method estimates the minimum required stiffness of the wall to restrict the lateral displacement within an acceptable value. The assumption is made that the wall

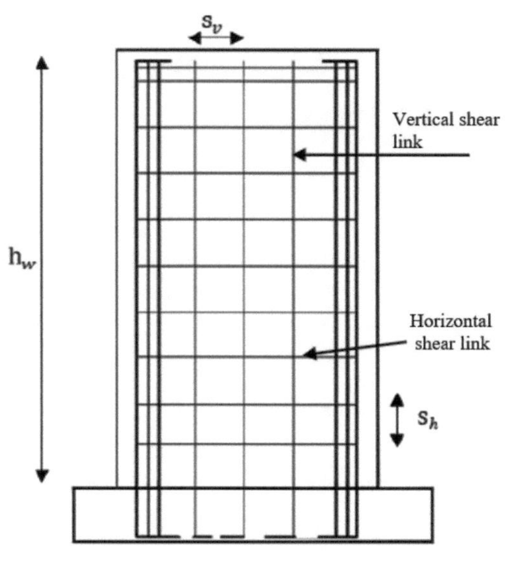

Figure 24.
Details of a typical shear wall.

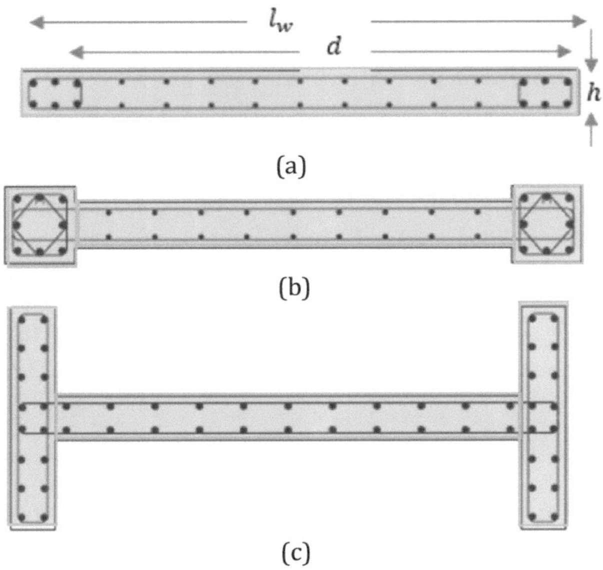

Figure 25.
Shear wall with boundary elements. (a) Boundary element inside the wall. (b) Shear wall with enlarged boundary element. (c) Shear wall with concentrated reinforcement in wings.

behaves like a cantilever beam subjected to a uniformly distributed load, with a constant rigidity of EI.

3.6.1 Wind load

To calculate the minimum uncracked moment of inertia for all walls in the direction of the wind load, considering a maximum drift of 1/500 of the story height, the following formula can be used [12]:

$$\sum I_g = \frac{500Vh_w^2}{2.8E_c} \tag{41}$$

where h_w is the wall height (**Figure 24**).

3.6.2 Seismic load

By employing a comparable approach, considering the code limit on the displacement, the minimum moment of inertia of the uncracked walls can be determined in the desired earthquake force direction [[2], Chapter 11]

$$\sum I_g = \frac{C_d Vh_w^2}{0.07E_c} \tag{42}$$

Once the minimum $\sum I_g$ is determined, the walls can be chosen in a manner that ensures the minimum moment of inertia, as specified in Eq. (42), is adequately provided.

$$I_{gi} = \sum I_g / n = \frac{hl_w^3}{12} \tag{43}$$

where h is the thickness and l_w is the length of the shear wall (**Figure 25**).

In the desired direction, the designer will assume the number of considered walls as 'n', and I_{gi} represents the second moment inertia of each wall. In accordance with the national regulations [13], the minimum thickness of the wall can be determined based on the shear resistance criterion. This determination of thickness of shear wall is determined assuming the maximum shear force that the wall can withstand, V_u, equal to maximum shear resistance of shear wall, along with a 50% safety factor

$$h \geq \frac{V_u}{0.5n\sqrt{f_c}d} \quad where\ h \geq \frac{h_w}{15} \tag{44}$$

For the initial design, d is assumed to be $0.8l_w$ ($d = 0.8l_w$) Replacing h in Eq. (43) by Eq. (44), the length of the each wall can be calculated as follows:

$$l_w = \sqrt{\frac{4.8\, n\sqrt{f_c}}{V_u} I_{gi}} \tag{45}$$

Usually, initially, I_{gi} of each wall is determined from Eq. (43), and then the initial length of the wall is calculated from Eq. (45), and finally, the initial thickness of each wall is obtained from Eq. (44). During the optimisation procedure, the dimensions of the wall can be changed.

Although it is deliberately simplified in the above calculations, it provides a close approximation to reality. It is clear that the final decision regarding the specifications of the walls will be made based on computer analyses and taking into account the effect of the frames and regulations. In addition, this method is also a strong and reliable tool for verifying the results of computer analysis.

Considering the size and height of a building, various shear walls are used (**Figure 25**). According to the ACI regulations [6], if the compressive stress caused by the joint effect of axial force and bending moment at the end of the compressive zone, under factored loads, is more than $0.2f_c$, the boundary elements (**Figure 25**) should be used. The relevant details are described in Chapter 14 [2]. The stress is calculated assuming that the walls are not cracked:

$$\sigma = \frac{P}{A} + \frac{M}{W} \geq 0.2f_c \tag{46}$$

Irrespective of the wall type, the procedure for determining the equivalent axial force remains consistent. The equivalent axial force acting on the boundary elements can be computed in the following manner (**Figure 26**).

$$\mathrm{P}_{ueq1} = \frac{P_u}{2} - \frac{M_u}{(0.8l_w)} \quad \text{Tensile force} \tag{47}$$

$$\mathrm{P}_{ueq2} = \frac{P_u}{2} + \frac{M_u}{(0.8l_w)} \quad \text{Compressive force}$$

The initial dimensions of boundary elements can be determined using the above equivalent axial force on the boundary elements

$$h_1 b_1 \text{ or } h_2 b_2 \geq \frac{\mathrm{P}_{ueq2}}{0.7f_c} \tag{48}$$

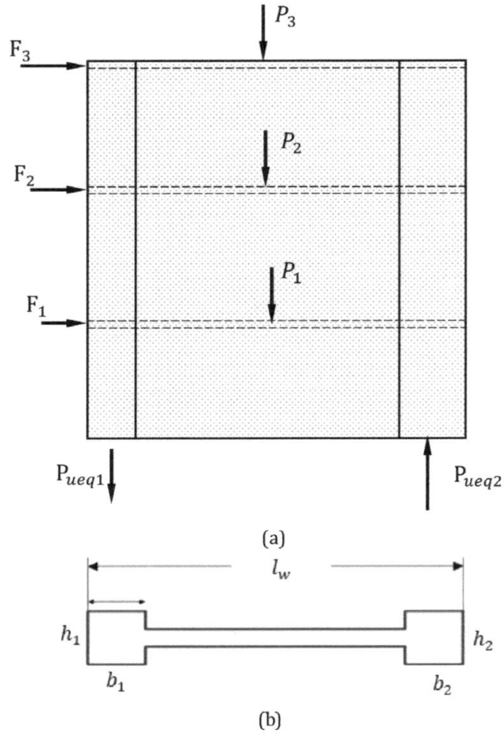

Figure 26.
Geometric characteristics of common shear walls for the simplified C and T method. (a) Equivalent axial force on boundary elements. (b) Cross section.

$$A_{st} \geq \frac{\mathrm{P}_{ueq1}}{0.9f_y} \tag{49}$$

A_{st} is the area of longitudinal rebars in the columns or in boundary elements.

3.7 Case study 3: Shear Wall subjected to lateral loads from earthquakes

In a five-story concrete structure featuring a flat-slab design, in axis X, three symmetrical shear walls are employed to resist lateral forces. As illustrated in **Figure 27**, the roof experiences a dead load of 6.5 kN/m^2 and a live load of 2.0 kN/m^2. The task involves designing a shear wall that incorporates a boundary element. At the base of the wall, the dead load is determined to be 220 kN, while the reduced live load is 50 kN. The concrete utilized has a compressive strength of $f_c = 25MPa$, and the reinforcing bars exhibit a tensile strength of f_y =400 MPa.

1. Estimation of the basic shear force. Using Eqs. (6) and (8) and assuming C = 0.165, the basic shear force would be

$$V = CW$$

$$W = nA(DL + 0.2LL + 4.5)1.3$$

$$W = 5(228)(6.5 + 0.2(2) + 4.5)1.3 = 16{,}894.8\ kN$$

$$V = 0.1586[16{,}894.8\ ] = 2680.44\ kN$$

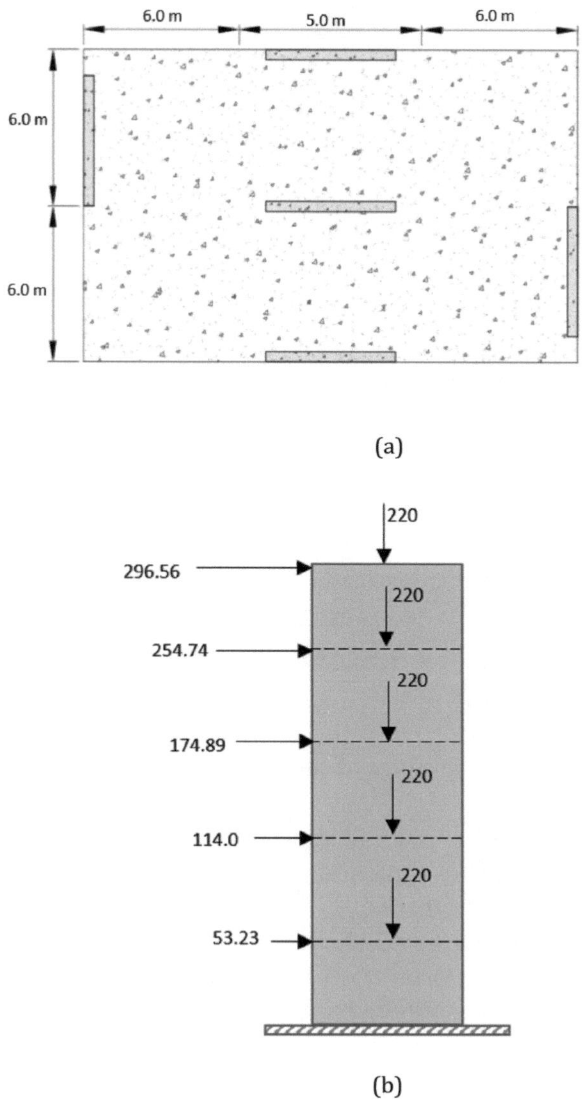

Figure 27.
Shear wall. (a) Plan. (b) Lateral load on each wall (kN): the height of ground floor is 2.8 m and the rest is 3.2 m.

2. Initial dimension

$$\sum I_g = \frac{C_d V h_w^2}{0.07 E_c} = \frac{5.5\,(2680.44)1000(15600)^2}{0.07(25000)} = 2.05E12$$

where

C_d= displacement magnification factor, e.g., $C_d = 5.5$

E = 25,000 MPa

Consequently, the moment of inertia for each wall is determined to be $I_{gi} = 2/05\text{E}12/3 = 6.83E11$. The wall length is derived from Eq. (45)

$$l_w = \sqrt{\frac{4.8\,n\sqrt{f_c}}{V_u} I_{gi}} = \sqrt{\frac{4.8(3)\sqrt{25}}{2680.44(1000)} 6.83E11} = 4283.25, l_w = 5000\ mm$$

To calculate the thickness Eq. (44) is used

$$h \geq \frac{V_u}{0.5n\sqrt{f_c}d} = \frac{2680.44(1000)}{0.5(3)\sqrt{25}(0.8)5000} = 89.35\ where\ h \geq \frac{h_w}{15} = \frac{2800}{15}$$
$$h = 186.67\ mm$$

The minimum thickness must be 300 mm, as the thickness of the boumadry elements is equivalent to that of the shear wall, so $h = 300\ mm$ and $l_w = 5000\ mm$.

According to Iranian code [13], to design shear walls, two load combinations, $1.2D + 1.0L + 1.0E$ and $0.9D + 1.0E$, are taken into account to calculate critical bending momnet and axial force on each wall (**Figure 24**).

$$\mathrm{M}_u = 1.0\sum P_i h_i = 1.0[296.56(15.6) + 254.74(12.4) + 174.89(9.2) + 114.06(6.0) + 53.23(2.8)] = 10227.5\ kNm$$

$$N_{u1} = 1.2[(220)5 + 0.3(5)15.6\ (25)] + 1.0(5)50 = 2272.0\ kN$$

$$N_{u2} = 0.9[(220)5 + 0.3(5)15.6\ (25)] = 1516.50\ kN$$

The maximum stress on the edge of shaer wall is

$$\sigma_c = \frac{N_{u2}}{hl_w} + \frac{6M_u}{hl_w^2} = \frac{2272.0\ (1000)}{300(5000)} + \frac{6(10227.5)10^6}{300(5000^2)} = 9.69\ MPa$$

As $\sigma_c = 9.69MPa$ is more than $0.2f_c = 5\ MPa$, the boundary elements needs to be provided. In this category of shear walls, the thickness of boundary elements is equivalent to the shear wall itself; therefore, the length of the boundary can be determined using Eq. (47)

$$\mathrm{P}_{ueq2} = \frac{N_{u2}}{2} + \frac{M_u}{(0.8l_w)} = \frac{2272.0}{2} + \frac{10227.5}{[0.8(5.0)]} = 3692.87kN$$

$$h_1 b_1 \text{ or } h_2 b_2 \geq \frac{\mathrm{P}_{ueq2}}{0.7f_c} = \frac{3692.87(1000)}{0.7(25)} = 211021.14\ \mathrm{mm}$$

$$b_1 = b_2 = \frac{211021.14}{300} = 703.41\ mm, b_1 = 750\ mm$$

The area of longitudinal reinforcement in boundary elements is determined using Eq. (49)

$$A_{st} \geq \frac{\mathrm{P}_{ueq1}}{0.9f_y}$$

$$\mathrm{P}_{ueq1} = \frac{N_{u1}}{2} - \frac{M_u}{(0.8l_w)} = \frac{1516.50}{2} - \frac{10227.5}{[0.8(5.0)]} = -1798.63kN$$

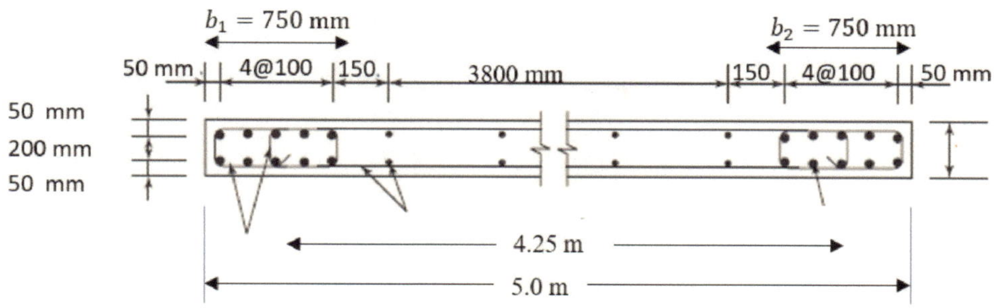

Figure 28.
Shear details.

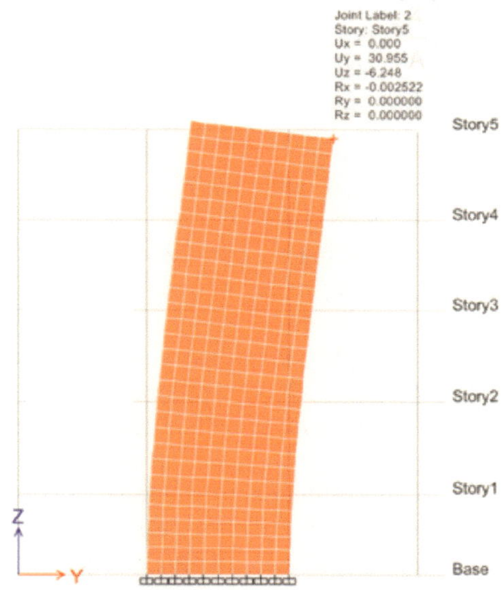

(a)

Flexural Design for P and M_3 —Tension Reinforcement

Station Location	Edge Length mm	Rebar Area mm²	Tension Combo	P_u kN	M_{u3} kN-m
Left Top	600	3122	DWal7	1264	7725.76
Right Top	600	3122	DWal8	1264	-7725.76
Left Bot	750	4812	DWal7	1348	10227.504
Right Bot	750	4812	DWal8	1348	-10227.504

(b)

Figure 29.
ETABS model. (a) Softwar model. (b) Results.

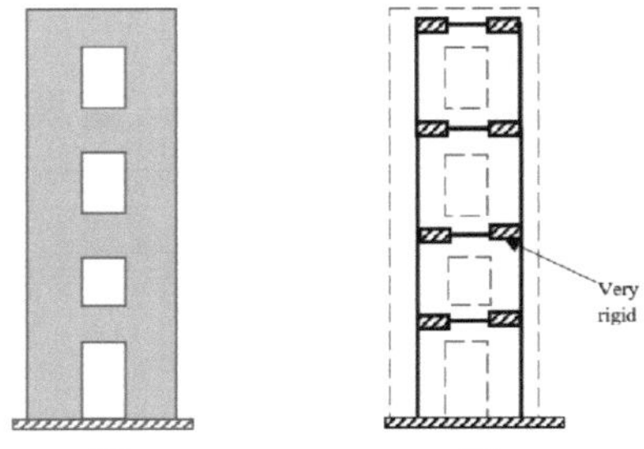

Figure 30.
Idealized coupled shear walls model. (a) Coupled shear wall (b) Idealized structural model.

$$A_{st} \geq \frac{1798.63(1000)}{0.9(400)} = 4996.19mm2\ use\ 10H25$$

To demonstrate the effectiveness of the proposed method, the identical project was simulated using ETABS. The findings indicate a strong correlation of the proposed method (**Figure 28**) with ETABS results (**Figure 29**) regarding the dimensions of the boundaries and the area of the longitudinal reinforcement bars.

3.8 Approximate analysis of coupled shear walls

Shear walls can be simplified by using an equivalent frame, as shown in **Figure 24**. In this frame, the column's moment of inertia matches that of the wall on each side of the opening. Similarly, the beam's moment of inertia matches the wall between the upper and lower openings. The moment of inertia of the beam in the rigid section, outside the openings, can be assumed as $100I_b$ (**Figure 30**).

Finite element method is commonly employed in the analysis and design of coupled walls in practice, thanks to the availability of reliable commercial software such as SAP2000 and ETABS. This eliminates the necessity for certain simplifications. Nevertheless, in order to comprehend the behavior and lay down a foundation for verifying the outcomes of computer analysis, an approximate analysis method is suggested here. This method is based on the results of software analysis carried out on buildings with 8, 12, and 16 stories by the author. Further information regarding these specific types of walls can be found in Chapter 14 [2].

The bending moment, M_o, and shear force, V_{si}, on each story can be determined as follows:

$$M_o = \sum F_i h_i \tag{50}$$

and

$$V_{si} = \sum F_i \tag{51}$$

The absence of a turning point at the same location as the story height in the coupling wall prevents manual calculation of the shear force, unlike in frames. The software analysis conducted by the author on the coupled walls [[2], chapter 11] reveals that, in a minimum of 50% of the stories, the turning point in the columns is situated very near the top of the story. As we progress upward, the turning point gradually shifts toward the middle of the story in the upper stories. Nevertheless, by assuming the turning point to be in the middle of the coupled beam, the shear forces up to the initial 50% of the stories can be determined conservatively

$$V_{bi} = V_{si}(0.8)h_w/(l_b) \tag{52}$$

Where V_{bi} is the beam's shear force, V_{si} is the shear force in the story, h_w is the story height, and l_b is the length of coupled beam.

For upper stories, the shear force of the coupling beam is calculated as follows:

$$V_{bi} = [(V_{si}h_{wi}/2 + V_{si+1}h_{wi+1}/2)]/(l_b) \tag{53}$$

where V_{si+1} is the shear force of the story and h_{wi+1} is the height of the column above the coupling beam.

The axial force – tensile or compressive – acting on the walls is obtained from the sum of the shearing force acting on the coupling beams in the floors above the desired column:

$$T_o = \sum_{i=1}^{n} V_{bi} = C_o \tag{54}$$

After calculating V_{bi}, the bending that is tolerated by the walls, $T_o l$, can be determined. Therefore, the contribution of the walls from the bending moment of M_o can be calculated [2]

$$M_{w1} + M_{w2} = M_o - T_o l_b \tag{55}$$

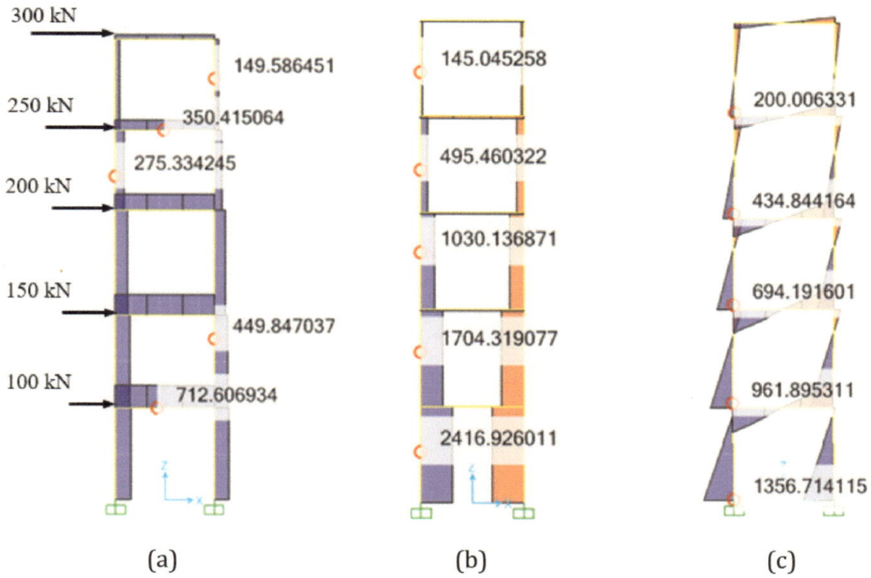

Figure 31.
Analysis of the coupled wall assuming the beam- column model similar to ***Figure 29b*** *using SAP 2000, span 3.2 m, floor height 3 m. (a) Lateral force and shear force diagram (b) Axial force diagram (c) Bending moment diagram.*

3.9 Case study 3: Coupled shear wall

To confirm the aforementioned relationships (Eqs. (52)-(54)), a coupled shear wall was simulated in SAP2000, as illustrated in **Figure 31b**. Based on Eq. (52), the shear force acting on the link beam of the first floor is equivalent to

$$V_{b1} = \frac{V_{s1}(0.8)h_{s1}}{l_b} = \frac{1000(0.8)3.0}{3.2} = 750\ kN$$

Also, the shear force acting on the link beam of the fourth floor is equivalent to

$$V_{b4} = [(V_{s4}h_{s4}/2 + V_{s5}h_{s5}/2)]/(l_b)$$

$$V_{b4} = \frac{[550(3)/2 + 300(3)/2]}{3.2} = 398.44\ kN$$

The results presented indicate that the shear forces experienced by the beams on the first and fourth floors closely align with the findings from SAP 2000, as illustrated in **Figure 31a**.

Using **Figure 31a**, we have

$$M_o = \sum F_i h_i = 10500\ kNm$$

$$T_o = \sum_{i=1}^{n} V_{bi} = C_o = 2416.93\ kN$$

The axial force of shear walls, as illustrated in **Figure 31a** and expressed in Eq. (54), aligns closely with the results obtained from the software, as shown in **Figure 31b**.

In this instance, the ratio $T_o l_b / M_o$ is determined to be 0.74, suggesting that the structure can be classified as a coupled beam [13]. Given that the moments M_o and $T_o l_b$ are specified in Eq. (55), it is possible to compute the moment contribution arising from the shear walls.

$$M_{w1} + M_{w2} = 10500 - 2416.93(3.2) = 2765.82\ kNm$$

As the stiffness of both walls is identical

$$M_{w1} = M_{w2} = 1381.91\ kNm$$

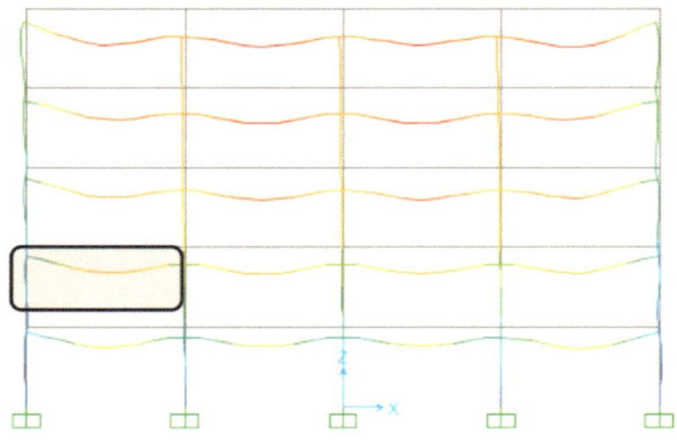

Figure 32.
Deflection due to gravity loads.

which agrees well with software result (**Figure 31c**)

The analysis of the proposed analytical method in conjunction with the outcomes from SAP2000 indicates that the method in question possesses adequate precision for the validation of the projects.

3.10 Approximate deflection

According to the developed method by author [2], in a continuous beam of a rigid frame (**Figure 32**), the deflection in each span can be determined using the following expression

$$y_{max} = \frac{5l^2}{48EI_e}[M_s - 0.1(M_2 + M_1)] \tag{56}$$

where

M_s is the positive bending moment in the middle of span length and M_1 *and* M_2 are the end bending moment. I_e is effective moment of inertia; $I_e = 0.5I_g$, and E is elastic module; $E = 5000\sqrt{f_c}.f_c$ is the compressive strength of concrete.

Considering the method of ACI coefficients for the analysis of beams in rigid frames, the maximum deformation in the side and middle spans is calculated as follows:

Side span lengths

$$M_1 = \frac{ql^2}{16} M_2 = \frac{ql^2}{10} \tag{57}$$

By substituting the above values into Eq. (56), the maximum deflection in the side spans is equal to

$$y_{max} = \frac{ql^4}{174EI_e} \tag{58}$$

Middle span lengths

$$M_1 = \frac{ql^2}{11} \qquad M_2 = \frac{ql^2}{11} \tag{59}$$

By substituting the above bending moments into Eq. (56), the maximum deflection in the middle spans is determined to be equal to

$$y_{max} = \frac{ql^4}{217EI_e} \tag{60}$$

Eqs. (58) and (60) can serve as effective expressions for validating computer modeling as well. If the computer-generated results for the vertical deflection of the beams or slabs closely match the aforementioned values or exhibit an acceptable deviation, along with the approximate structural analyses, it can be inferred that the loading data and computer modeling have been verified.

3.11 Lateral displacement

The previous section covers the analysis of deflection in continuous beams. Various deformations must be considered when designing reinforced concrete frames.

One of the most significant deformations is the lateral displacement of the frames. It is crucial to control this displacement to avoid causing discomfort to residents and damage to partitions, facades, and windows. As lateral displacement can result in an increase in the P-Δ effect, which is inversely proportional to the lateral stiffness, controlling it becomes crucial. Typically, software is used to calculate the amount of lateral displacements but manual calculations is crucial for software structural and analysis verifications.

The horizontal displacement of each floor level can be determined based on the shear force and stiffness of the structure. If we consider the floors to be rigid, the lateral stiffness of the frames on each story can be computed in the following manner [8]:

$$k = n\frac{12EI}{h^3} \tag{61}$$

Based on the research conducted by the author [2], it has been found that the stiffness mentioned above is significantly greater than the true value. Consequently, according to analytical calculation and software modeling, the author suggests a modified stiffness that takes into account the beams when calculating the stiffness of the frames.

$$k = n\frac{5EI}{h^3}\ \text{Ground floor} \tag{62}$$

$$k = n\frac{3EI}{h^3}\quad \text{Other floors} \tag{63}$$

Upon determining the shear force and story stiffness, the relative displacement of each story, denoted as Δ, can be calculated in the subsequent manner.

Where

Δ_{rei} is relative displacements

V_i is shear force in the story

K_i is story stiffness

$$\Delta_{rei} = \frac{V_i}{k} \tag{64}$$

The procedure to calculate lateral displacement in a five-story building is described in the **Table 6**. The method can be applied to buildings with various stories.

Story	Lateral force (kN)	Shear force (kN) V_i	Story stiffness, N/mm	Relative displacement $\Delta_{rei} = \frac{V}{k}$ (mm)	Lateral displacement (mm) $\Delta_i = \sum \Delta_{rei}$
4	F4	F4	K4	V4/K4	$\Delta_{re0} + \Delta_{re1} + \Delta_{re2} + \Delta_{re3} + \Delta_{re4}$
3	F3	F3 + F4	K3	V3/K3	$\Delta_{re0} + \Delta_{re1} + \Delta_{re2} + \Delta_{re3}$
2	F2	F2 + F3 + F4	K2	V3/K2	$\Delta_{re0} + \Delta_{re1} + \Delta_{re2}$
1	F1	F4 + F3 + F2 + F1	K1	V1/K1	$\Delta_{re0} + \Delta_{re1}$
Ground	F0	F0 + F1 + F2 + F3 + F4	K0	V0/K0	Δ_{re0}

Table 6.
Lateral deflection (Example for five stories).

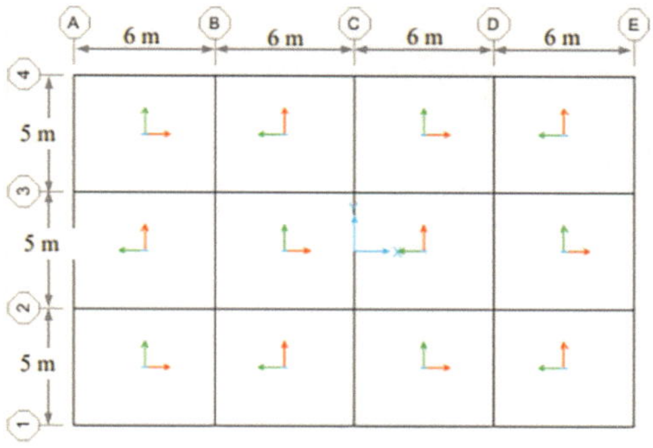

Figure 33.
Floor plan.

3.12 Case study 4: Vertical and lateral deflection

In a five-story educational building located in Sanadaj, Kurdistan, a joist system has been utilized for the roofing framework (see **Figure 33**). The following is a summary of the loading conditions:

$$\text{DL} = 4.95\, kN/m^2$$

$$\text{LL} = 3.5\, kN/m^2$$

It is estimated that 20% of the live load may be regarded as a permanent load. In addition, to avoid damaging the partitions, a gap of about 35 mm between the beams and the partitions is filled with foam.

3.12.1 Approximate analysis and design

3.12.1.1 Initial design

In accordance with the methodology outlined in case study 1, the preliminary demotions of the beam and column are shown in **Table 7**.

3.12.1.2 Vertical deflection in the beams

According to the proposed approximate method, deflection resulting from the permanent load, $\delta_{i,sus}$ in beam 2 (**Figure 33**), can be calculated. Given that the lengths of the spans are identical, the maximum deflection will take place in the side spans

$$\delta_{i,sus} = \frac{q_{sus} l^4}{174 E I_e} \text{ Side spans}$$

Story	4	2,3	G, 1	
Columns	450(450)	400(400)	350(350)	mm
Beam	350(500)	350(450)	300(400)	mm

Table 7.
Summary of initial design.

Where
The loading width for beam 2 is 2.5 m, then

$$q_{sus} = 2.5(4.95 + 0.2(3.5)) = 14.13\ kN/m$$

$$I_e = 0.5I_g = 0.5\frac{bh^3}{12} = 0.5\frac{350(500)^3}{12} = 1.82\left(10^9\right)\ mm^4$$

$$E_c = 5000\sqrt{f_c} = 5000\sqrt{25} = 25000\ MPa$$

$$\delta_{i,sus} = \frac{14.13(6000)^4}{174(25000)1.82\left(10^9\right)} = 2.31\ mm$$

The same structure is modeled by SAP 2000. The comparison between the outcomes of manual calculations and software analyses demonstrates the effectiveness of the proposed approximate method. As illustrated in **Figure 34**, the immediate

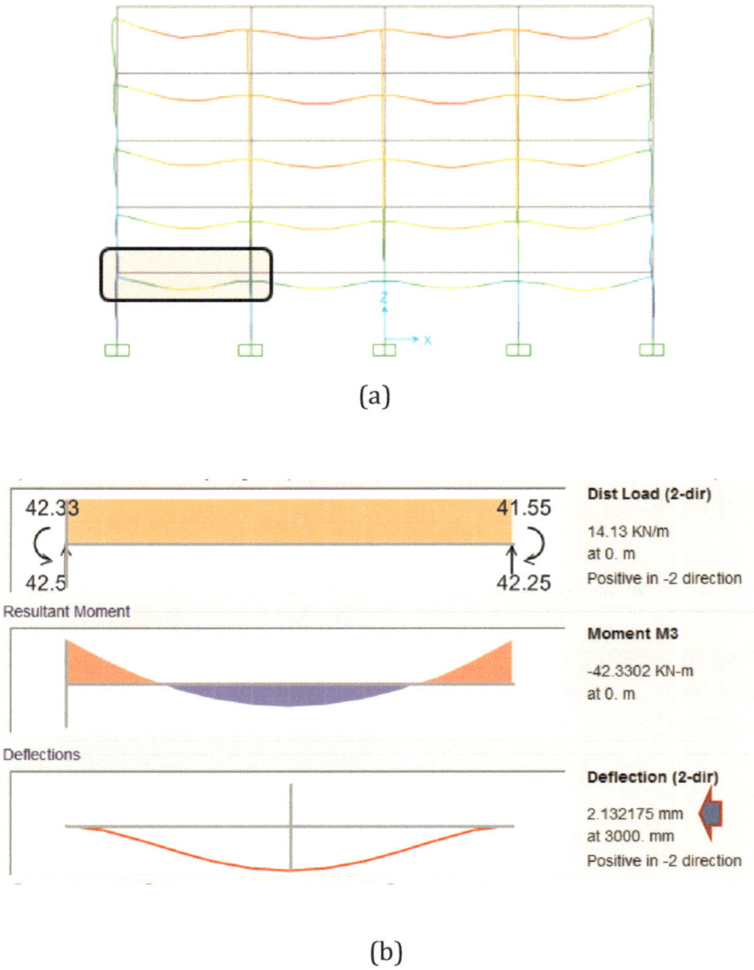

Figure 34.
Immediate deflection – software analysis. (a) Deflection under dead load – beam 2. (b) Bending moment diagram and deflection under dead load.

deflection of beam2 derived from software analysis closely aligns with the results obtained from the approximate approach, denoted as $\delta_{i,sus}$.

3.12.1.3 Lateral displacements

To determine the relative displacement of various stories, as indicated in Eq. (64), it is essential to consider the shear force along with the corresponding stiffness values. The lateral forces exerted on the structure are depicted in **Figure 35**, while the stiffness of the individual stories is defined by Eqs. (62) and (63). The method outlined in **Table 6** is employed to estimate the lateral displacement (**Table 8**).

Story stiffens

$$k = n\frac{5EI}{h^3} = 20\frac{5(23500)450^4/12}{3000^3} = 297{,}421\ N/mm \text{ Ground floor}$$

$$k = n\frac{3EI}{h^3} = 20\frac{3(23500)450^4/12}{3000^3} = 178{,}453\ N/mm \text{ Other stories}$$

Additionally, an examination of the lateral displacement presented in **Table 6**, when juxtaposed with the software analyses depicted in **Figures 36** and **37**, reveals that the discrepancy is minimal, measuring less than 7%.

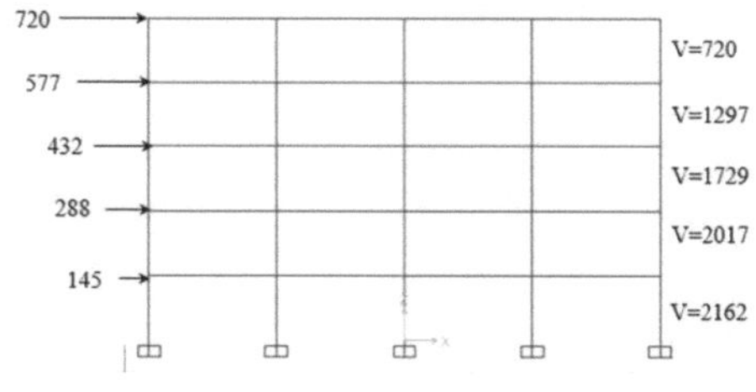

Figure 35.
Lateral load on the building and story shear force.

Story	Lateral force (kN)	Shear force (kN)	Story stifness (N/mm)	Relative displacement $\varDelta = \frac{V}{k}$ (mm)	Lateral displacement (mm)
4	720	720	178,453	4.04	39.55
3	577	1297	178,453	7.27	35.51
2	432	1729	178,453	9.67	28.24
1	288	2017	178,453	11.30	18.57
G	145	2162	297,421	7.27	7.27

Table 8.
Lateral displacements- proposed method.

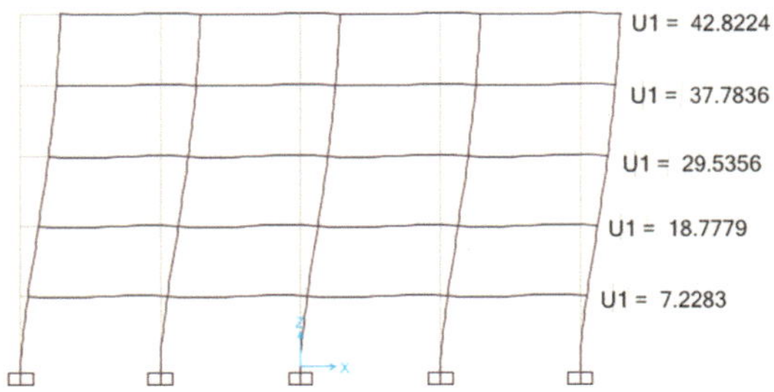

Figure 36.
Lateral displacement – software analysis.

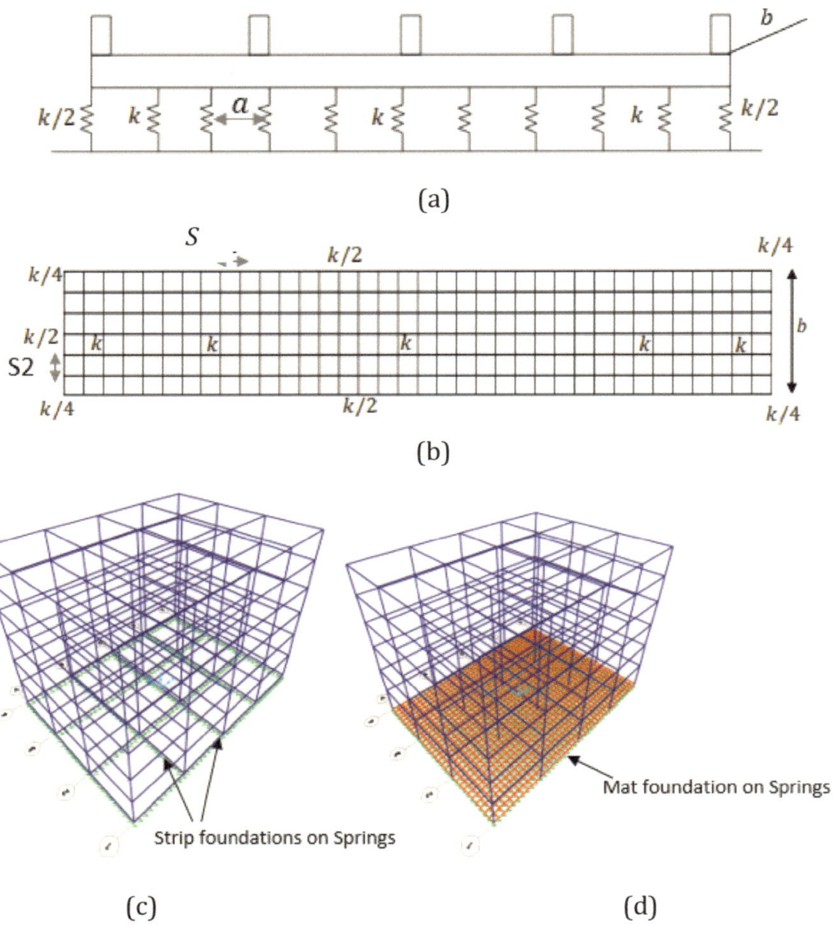

Figure 37.
Approximate analysis of foundations – modeling soil by springs. (a) Beam model; $k = baK_s$ *(b) Shell model;* $k = S_1S_2K_s$ *(c) Using beam element and spring to model strip foundation (d) Using shell element and spring to model mat foundation.*

3.13 Approximate analysis of foundations

SAFE software is utilized for the analysis and design of foundations. This software employs finite element (FE) methods and springs to accurately represent the behavior of the soil. However, due to the complex two-way interaction between the foundation and the soil, manually verifying the analysis of foundations would pose considerable challenges. In order to address this issue and offer a practical verification approach, two methods can be employed depending on the type of foundations being considered.

3.13.1 Two-way strip foundations

To validate the analyses and design conducted using SAFE, it is essential to follow the outlined procedures.

- The primary software utilized for modeling the frame is employed.
- Supports at the base of the columns are eliminated.
- Beam elements, matching the dimensions of the foundations, are used to connect the columns (**Figure 37c**).
- The foundations are segmented, and springs are allocated to the joints (**Figure 37c**). The properties of these springs can be determined based on soil stiffness (K_s) as indicated in the soil report.
- Both the structure and foundation are analyzed within a single model.
- A comparison is made between the bending moment results obtained from this method and those from SAFE, thereby confirming the accuracy of the SAFE results.

3.13.2 Mat foundations

1. *Software*: The previously discussed method can also be utilized in this context. Specifically, for modeling the foundation, a combination of a shell element and a spring is employed (see **Figure 37d**).

2. *Analytical*: Based on the author's experience, employing approximate analysis for flat slabs subjected to gravity loads can achieve an adequate level of accuracy. Each strip (**Figure 38**) may be evaluated utilizing a uniformly distributed load (UDL) of Q_u, along with the coefficient method as outlined in **Tables 9** and **10**.

$$Q_u = \frac{\sum R_{ui}}{bl}$$

Where
R_u – factored reaction supports
b and l are the foundation's dimensions.
It is important to acknowledge that this method is applicable solely to gravity load.
To calculate the thickness of pad, strip, combined, and mat foundation, two methods can be utilized:

1. Thickness of foundation is two times of column dimension

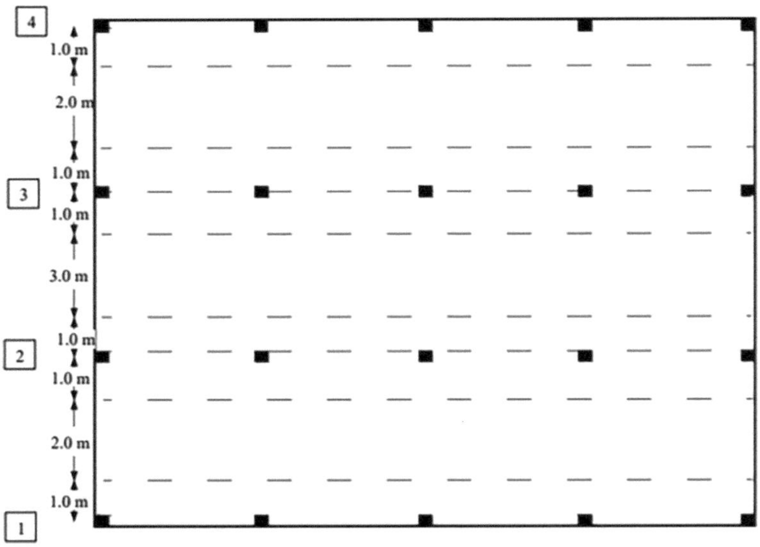

Figure 38.
Mat foundation.

		at the first internal column		*at the middle of span length*	*at the middle columns*	
M_u	α	−0.086		0.063	−0.063	
V_u	β	0.6	0.6	—	0.5	0.5

$M_u = \alpha q_u l_n^2$.
$V_u = \beta q_u l_n$.
$q_u = (l_x R\, l_y) Q_u$.

Table 9.
Approximate analysis of mat foundations under gravity load [4].

	Column strip	Middle strip
$M_{uc} = k_{ci} M_u$	$k_c^- = 60–80\%(70\%)$	$k_c^+ = 20–40\%(30\%)$
$M_{um} = k_{mi} M_u$	$k_m^- = 50–65\%(55\%)$	$k_m^+ = 30–50\%(45\%)$

Table 10.
Dividing the bending moment between the middle and column strips [2].

2. Simplified analytical method proposed by author [[2], Chapter 12]

$$(2.6v_c)d^2 + [1.3(c_1 + c_2)v_c]d - P_u = 0 \quad (65)$$

Where
v_c is the two-way shear strength of the concrete
c_1 *and* c_2 is column dimension
P_u maximum factored axial force in the columns
d is effective depth of foundation, so foundation depth is $h = d + 75\,mm$

4. Conclusion

The advanced software available for structural analysis and design is widely utilized by consulting engineers and professionals within the engineering field; however, students typically do not engage with such software during their academic training. It is crucial to emphasize that the application of this software is reserved for engineers who possess adequate knowledge and skills in manual analysis and design. Consequently, a computer cannot supplant the role of an engineer under any circumstances. The primary function of computer software in this context is to enhance efficiency and streamline processes. The relationship between engineers and computers can be likened to that of a surgeon and a scalpel; the effectiveness of the tool is contingent upon the expertise of the individual wielding it. Ultimately, while both the computer and the scalpel serve as valuable instruments, it is the trained professional who is capable of achieving meaningful outcomes. This chapter book addresses and rectifies the aforementioned limitations through a scientific approach.

The newly developed approximate analysis and design methodology for reinforced concrete (RC) structural elements, validated through four case studies, offers a robust foundation for structural designers to verify software modeling, analysis, and design processes for complex projects. Additionally, this approach will assist students in estimating internal forces and dimensions of various elements within multi-story RC structures, particularly in scenarios where traditional analytical methods may fall short.

5. Questions

1. Explain how the material for a specific building is chosen?
2. For a concrete building, what factors are effective in choosing the gravity and lateral bearing system?
3. What is the main difference between tall and low-rise buildings?
4. Deformation of a 12-story concrete building is more than the permissible limit. If only the bending frame is used, increasing the dimensions of the column or beam provides a more economical result, why?
5. In a composite system, the flexural frame alone must sustain 30% of the lateral load alone. Meanwhile, shear walls should also be designed for 100% lateral load. It seems that 130% is considered in the design. Explain the relevant concept?
6. Why does the dual system's efficiency decrease as the building height increases?
7. Tips to enhance the performance of a flexural frame against lateral load?
8. What is the maximum allowable height for a rigid frame with medium ductility?
9. Define tall building?
10. Illustrate the tubular system by sketching the diagram.
11. Explain the strengths and weaknesses of tubular systems.
12. Discuss the basic philosophy of using tubular systems.
13. What is the economical height for tubular systems?
14. What is the main philosophy of using bundled tubes and tube-in-tube system?
15. Find the best position for outrigger and belt trusses along the height of an 80-story building. Consider the use of one, two, three, or four outrigger and belt trusses.
16. What criteria are used to select gravity load-bearing systems for floors/roofs?
17. Consider three concrete buildings with 5 m, 10 m, and 15 m center-to-center columns. Suggest the most economical roofing system for the buildings. Discuss your choices.
18. The width of a 15-story concrete building is 10 meters, while the width of a 30-story concrete building is 42 meters, which one can be considered a tall building, why?

6. Problems

1. The plan of a seven-story concrete building is shown in **Figure 39**. The building is located in the city of Sanandaj. The building utilizes a medium gravity and lateral load bearing system within its rigid frame. By adjusting the dimensions of columns and beams every two stories, along with selecting the suitable floor system, the building can be effectively designed using the methods outlined in this chapter. The designer has the flexibility to make other necessary assumptions.

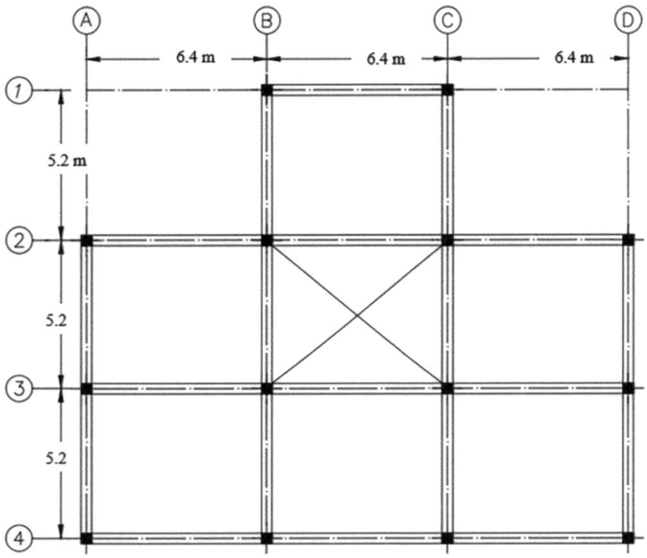

Figure 39.
Floor plan – medium bending frame.

2. The design of a 19-story concrete building in Mahabad is illustrated in **Figure 40**. The structure features a medium bending frame for gravity and lateral support, along with a special shear wall. Utilize the methods outlined in this chapter for approximate analysis and design, ensuring all assumptions are carefully selected by the designer.

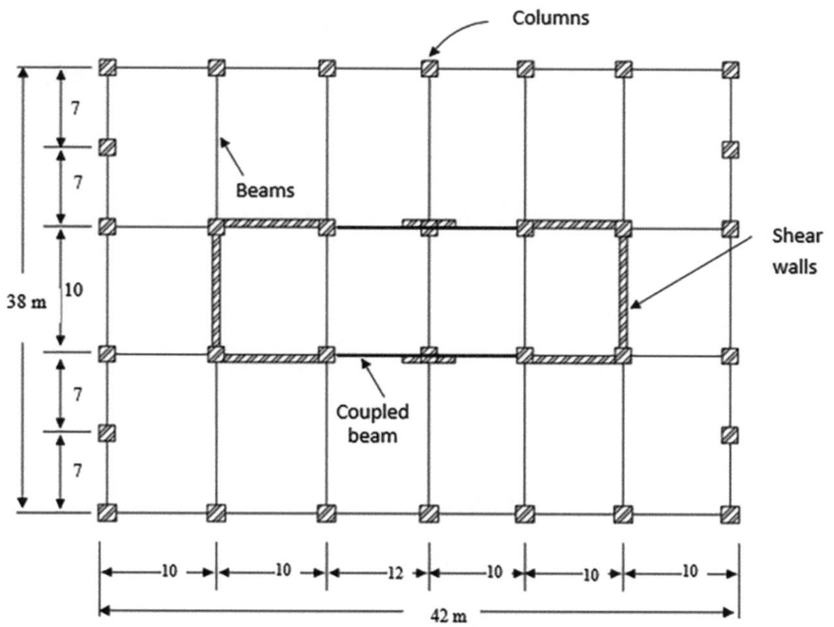

Figure 40.
Floor plan – dual load bearing system.

Author details

Mosleh Tohidi* and Ali Bahadori-Jahromi
School of Computing and Engineering, University of West London, Ealing, London, United Kingdom

*Address all correspondence to: mosleh.tohidi@uwl.ac.uk

References

[1] Tohidi M. Design of Reinforced Concrete Structures. Sanandaj Branch: The AZAD University; 2009. ISBN: 978-964-223-420-2

[2] Tohidi M. Advanced Design of Reinforced Concrete Structures.2009. Under publication, The Academic Center for Education, Culture and Research (ACECR). Iran; Tehran University

[3] Nilson A, Darwin D, Dolan CW. Design of Concrete Structures. Boston, USA: Mc Graw Hill; 2004

[4] Mosely B, Bungey J, Hulse R. Reinforced Concrete Design. New York, USA: Palgrave Macmilan; 2009

[5] BS EN1991-1-7. Action on Structures - Part 1–7: General Actions- Accidental Action. London, UK; 2006

[6] American Concrete Institute (ACI) Committee.318. Building Code Requirements for Structural Concrete (ACI 318-95) and Commentary (ACI318R-95). Detroit, MI, USA: ACI; 1995-2008

[7] Haukaas T. Approximate Methods. Vancouver: The University of British Columbia; 2015

[8] Podder D. Chatterjee D Approximate Analysis of Statically Indeterminate Structures. Boca Raton: CRC Press; 2021

[9] Leet K, Bernal D. Reinforced Concrete Design. Boston, USA: Mc Graw Hill; 1996

[10] Spencer W Jr. The Approximate Analysis of Structures. Springer; 2023. Kink

[11] Norris CH, Wilbur JB. Elementary Structural Analysis. 2nd ed. New York: McGraw-Hill; 1960

[12] Wight JK. Reinforced Concrete: Mechanics and Design. 6th ed. New Jersey, USA: Pearson; 2012

[13] Iranian National Building Regulation. Building Research center. Ministry of Housing. 2012

Chapter 7

Seismic Retrofitting Reinforced Concrete Structures with Precast Pre-Stressed Concrete Braces: An Overview

Mohsen Vatandoost and Fariborz Nateghi-A

Abstract

It is crucial to innovate retrofitting solutions for safeguarding buildings from catastrophic failure in earthquake-prone regions. An approach to retrofitting reinforced concrete (RC) buildings involves the use of precast pre-stressed concrete braces (PPCB). Employing PPCB for seismic retrofitting of RC structures is a novel method that has yet to be extensively researched. This approach obviates the need for wet concrete work at the site, thereby enhancing efficiency and minimizing the disturbance of the retrofitting process. This study aims to evaluate and develop this method. In this research, we investigated the seismic performance of this system using finite element analysis (FEA). The results showed that PPCB effectively decreased lateral displacement, changing the force-resisting mechanism into truss action and improving the building's seismic performance. However, when the brace's compression strength is lower than the existing frame's, the retrofitting system exhibits low stiffness and is ineffective at reducing lateral drift. Conversely, in models with two or more compressive strength ratios (brace to frame), braces demonstrated high strength and stiffness. Thus, PPCB can provide significant strength and stability to structures. Furthermore, we proposed and evaluated two novel configurations for this retrofitting system: a single diagonal and a V-shape.

Keywords: seismic retrofitting, seismic strength, concrete braces, precast concrete, pre-stressed concrete, reinforced concrete structures, finite element modeling, finite element analysis, single diagonal braces

1. Introduction

Reinforced concrete (RC) buildings in earthquake-prone regions must have sufficient strength, stiffness, and ductility to perform satisfactorily in a major earthquake. Seismic strength and retrofitting of existing buildings are essential due to the potential risks posed by earthquakes. Seismic retrofitting and strengthening existing buildings are important because they enhance the structure's ability to withstand earthquakes and reduce the risk of structural damage or collapse. This helps protect lives, property, and infrastructure during seismic events [1–3]. Retrofitting is often more cost-effective than rebuilding from scratch. It extends the lifespan of existing structures and avoids the excessive costs associated with reconstruction [4, 5]. The seismic strength and retrofitting of exciting buildings are inevitable due to changes in occupancy or usage, modifications to the applied loads on the building, adding floors, amendments in requirements in the building codes and regulations, etc. [4].

There are several methods and techniques to enhance the seismic performances of the buildings [6–9], including adding steel bracing [10, 11], adding concrete or steel shear walls [12], adding infilled masonry walls [13, 14], installing dampers [15–17] or base isolators [18, 19], reinforcing structural elements [9], upgrading connections [20], steel jacketing RC beam or columns [21], adding steel plates to improve beam-column connections [22], strengthening an RC frame by adding internal steel frame [23], and so on. However, selecting methods and techniques to be used for seismic retrofitting depends on various factors, including building type and design, structural assessment, budget and cost-effectiveness, preservation considerations, seismic hazards and locations, local building codes, etc. [24, 25].

Watanabe and colleagues [26] introduced a concrete brace for seismic retrofitting (**Figure 1**). The benefits of the proposed system included that there is no need to work with wet concrete at the site, no anchorage or bolts are required to the exciting frame, the installation is easy, and this system provides adequate stiffness for the lateral and seismic loads [26]. In addition, utilizing this system, building usage will continue and will not be disturbed; therefore, retrofitting costs will be reduced accordingly.

By using this proposed PPCB, the openings are not covered, and light and ventilation still exists, while in contrast, in some retrofitting methods such as adding concrete shear walls, the openings will be covered (see **Figure 2**).

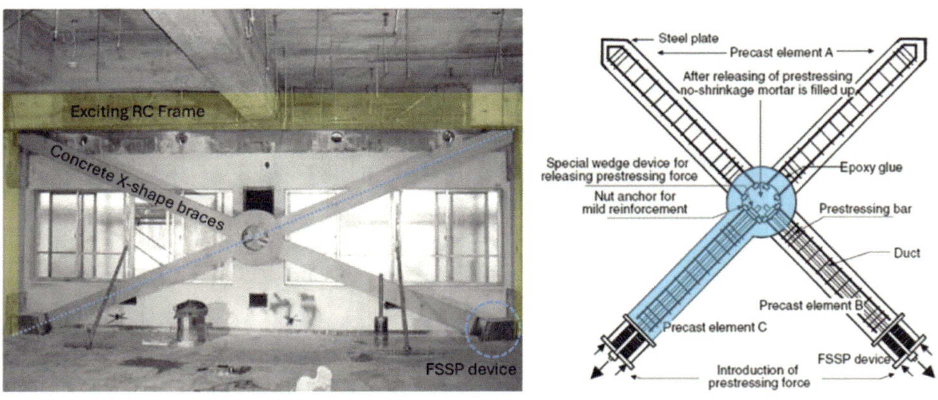

Figure 1.
The installation of concrete braces on an existing frame ensures that the openings (windows) remain unobstructed, thereby maintaining the flow of fresh air and light (a). The details and components of the retrofitting system (b), modified from Ref. [26].

The precast pre-stressed concrete braces (PPCB) consist of four concrete legs that are pre-cast and assembled on-site. One leg is designed with an extended part to provide a connection end, or the middle section (element C in **Figure 1b**, highlighted), which lets all these legs connect to shape an X-brace. The middle section will be filled with non-shrinkage concrete after assembly. To prevent squeeze caused by compression force between brace and existing frame, the X-shape brace is connected to the existing RC frame with a thick steel plate [26].

When the lateral loads (seismic or wind loads) act on the structure, one of the diagonal members of the X-shape brace goes on compression, and another is acting in tension [27]. Because concrete is not able to withstand tensile forces effectively, and due to its low tensile strength, which is about 10% of its compression strength [28], a special device (FSSP) is designed to let the concrete brace be elongated when it undergoes tensile forces. The flat springs and steel pipe (FSSP) device is depicted in **Figure 3**. In addition, the FSSP device ensures that braces are always active, even if they go under tensile force, the springs are elongated, and the force will transfer between brace and frame. The spring stiffness in the Watanabe test [26] was 2.5 KN/mm.

In the following, this method for seismic retrofitting of existing buildings using precast pre-stressed concrete braces (PPCB) is investigated, evaluated, and developed by the authors [5, 29–31]. In our research, the seismic performance of these braces was studied through finite element analysis (FEA)—using the ABAQUS software. Moreover, two novel configurations (single diagonal and V-shape) for this system were proposed, and further evaluated.

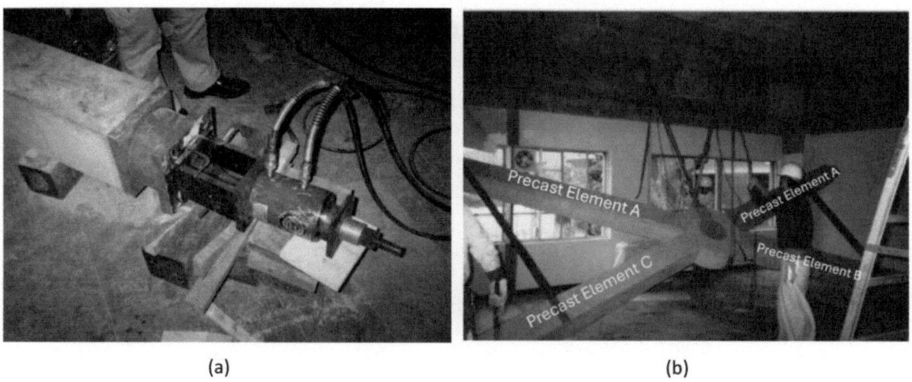

Figure 2.
Precast pre-stressed X-shape concrete braces; The braces are assembled, and tensile force is applied to pre-stress them (a); The braces are elevated in order to be positioned inside the current frame (b), modified from Ref. [26].

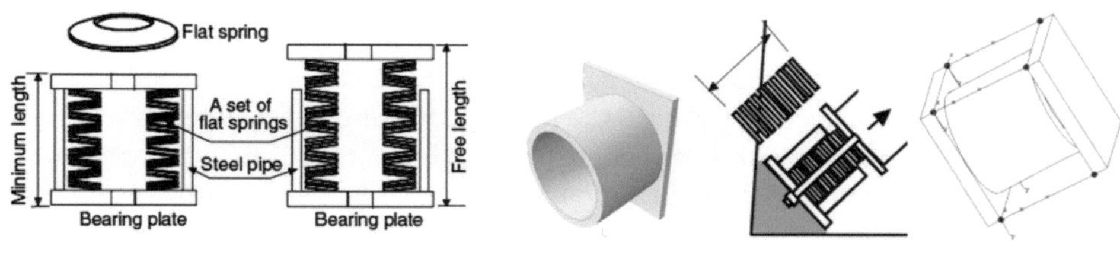

Figure 3.
The flat springs and steel pipe (FSSP) mechanism enables the concrete bracing to elongate when subjected to a tensile force, while guaranteeing that the concrete remains free from any tensile stress, modified from Ref. [26].

2. Background

The seismic retrofitting of structures in seismically hazardous regions is reflected in several studies, which indicates the importance of this topic. Vishnuprakash et al. [32] studied the behavior of various types of pre-stressed flat slabs under seismic effects in a single-story frame with and without braces. In a paper by Borah and Choudhury [33], which is presented at ASMA 2021 and is part of the book series STIN, a comparison based on different methods is presented. In addition, Lee et al. [34] proposed a new diamond-shaped bracing system with a horizontally layered friction damper for seismically strong structures. In their case study, they implemented it in a two-story R/C building, and the results indicated the effectiveness of the proposed method. Marrazzo et al. [18] employed mass dampers to enhance the structural performance. Similarly, Zhang et al. [35] investigated a base-isolated system. In another study conducted by Chong et al. [36], an experimental test was performed to evaluate the utilization of steel jacketing in RC frames. In a recent study by researchers from Taiwan [37], precast ultra-high-performance concrete braces were used for seismic retrofitting. With these braces, the force-resisting mechanism was transformed into truss action, and the seismic performance of RC frames was enhanced. Also, Maheri et al. [38] used various retrofitting techniques for retrofitting a school. Furthermore, Hung et al. [39] studied various methods for enhancing RC beam-column joints using engineered cementitious composite (ECC), fiber-reinforced polymer (FRP), and the concrete jacketing method. Cao et al. [40] proposed using new external sub-structures to retrofit the existing reinforced concrete frames.

Based on the provided literature, although numerous methods for seismic retrofitting of RC structures are used, employing precast pre-stressed concrete braces (PPCB) for seismic retrofitting RC structures is not reflected in much research and is an innovative method to be utilized in retrofitting. Based on this, the aim of this research was to study, evaluate, and develop the method of seismic retrofitting existing buildings by employing precast pre-stressed concrete braces (PPCB). In this research, we have investigated the seismic performance of these concrete braces by finite element analysis (FEA).

3. Method

To evaluate the seismic performance of the strengthened structure with precast pre-stressed concrete braces, we have created a model consisting of an existing frame and concrete braces based on the existing data from an experimental test by Watanabe [26], using ABAQUS (nonlinear-finite element) software (**Figure 4**) [5, 29–31].

Figure 4 illustrates that the finite element model was meticulously constructed to include all the essential components of the experimental model. The structure includes primary longitudinal rebars and shear reinforcements in both the existing frame and the brace. Furthermore, the model must take into account the mechanical properties and the connection between the longitudinal and lateral bars, as well as the bond between the concrete and steel, in order to ensure accurate transmission of forces between the steel bars and the concrete frame. The longitudinal rebars and shear rebars were 'Merged' in the model to create a single object and 'Embedded' in the concrete (beam or column).

Furthermore, as previously mentioned, pre-stress cables were utilized to replicate the gravitational load on the frame. To accurately replicate the experimental work,

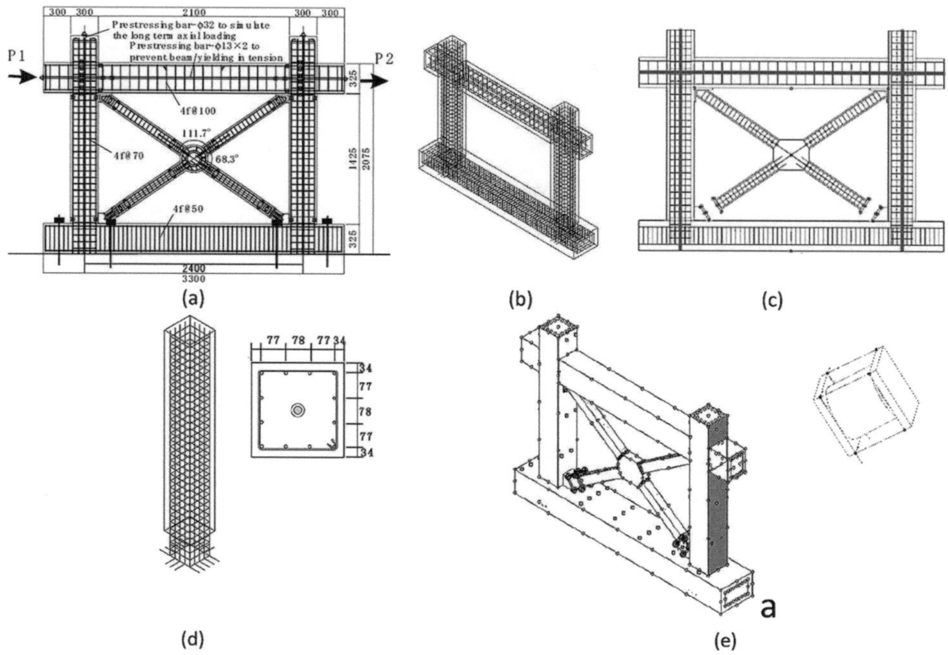

Figure 4.
Details of the X-shape precast pre-stressed concrete braces in the experimental work (a) and in the finite element model (b-e).

the connection between the pre-stressed cables and concrete was designated as an 'Unbounded Pre-stressing' steel bar in the model. Furthermore, the braces will always undergo compression, even though the tensile force is applied to the brace because of the FSSP device; therefore, the forces will travel between brace parts (leg 1 and leg 2) based on the pressure contact of the surfaces; this character is allocated in the FE model. Additionally, the FSSP device was simplified to two plates (170 mm × 140 mm × 20 mm) that were placed at a 140 mm distance from each other, and the corners of the plates were connected with four similar springs with a cumulative stiffness of 2.5 KN/mm, as demonstrated in **Figure 4(e)**.

In addition, the finite element (FE) model includes a steel endplate represented as a '3D element,' and the high-strength, non-shrinkage mortar is represented as a '3D-deformable element' in the FE model, as shown in **Figure 4**. Moreover, the foundation beam was modeled according to experimental work. All the frame and brace dimensions and properties are listed in **Tables 1–3**.

A 'tie' condition has been implemented on the contact surface between concrete parts of the frame (for instance, the cantilever beam and the column in the model) to ensure that all the concrete surfaces of the frame components are connected. This condition ensures that the movement and displacement of the second surface (the

Case	Cross section (mm)	Anticipated failure mode
No. 1	100 × 120	Buckling of the brace
No. 2	120 × 150	Shear failure at the end of a column or beam

Table 1.
Frame and brace cross sections and anticipated failure modes.

Case		f_c' (MPa)	E_c (GPa)	F_t (MPa)
No.1	Frame	27.5	26.1	2.53
	Brace	24.9	23.3	2.58
No.2	Frame	25	24.3	2.51
	Brace	64.1	33.3	4.52

Table 2.
Material properties of the braces and frame.

Compressive strength (f_c)	27.5 MPa (~275 kg/cm^2)
Tensile strength	2.75 MPa
Modulus of Elasticity (E)	26,220 MPa
Density	2400 kg/m3
Poisson's ratio	0.2
Shear strength	6 MPa

Table 3.
Base frame concrete properties.

slave) are connected to the primary surface (the master). Additionally, 'ties' are employed at the intersection of the columns with the foundation and the contact surfaces of the middle beam to the frame columns.

The model incorporated the boundary conditions during the 'initial' phase, followed by the application of loads in the subsequent step using a 'smooth step' approach to ensure a gradual application of the load rather than an abrupt impact load.

The loading setup that was employed in the experimental work is depicted in **Figure 5**. The cyclic load was employed with two hydraulic jacks, and the related horizontal displacement was measured throughout the test within the indicated points on the frame, which is shown in **Figure 5**. Moreover, to simulate the gravity loads of the above story on the frame, the columns were pre-stressed through the cables. In addition, the top beam in the frame was pre-stressed to prevent tensile failure. The same loading strategies were employed in the simulation (refer to [29]). The loading

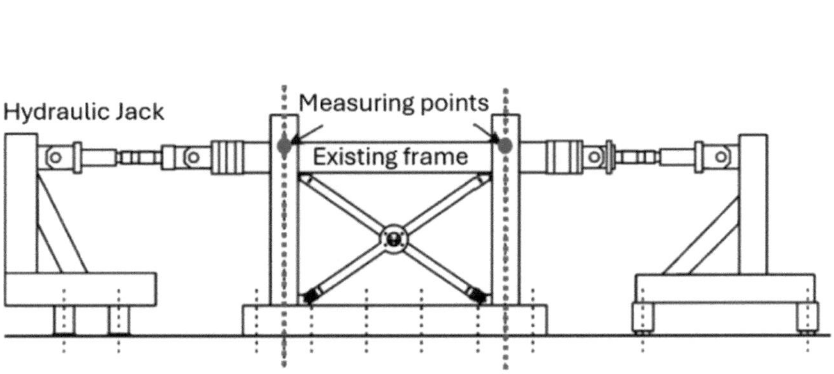

Figure 5.
Loading setup and measuring point for lateral displacement. The cyclic load was applied with two hydraulic jacks.

sequence (steps) was as follows: post-tension loads on the columns and beam, then applying the lateral loads through the jacks in the incremental steps.

First, we compared the results of the experimental study with the finite element model to ensure the accuracy of the finite element model. Watanabe et al. created and assessed half-scale concrete braces. The frame and brace material properties, and the cross sections are listed in **Tables 1–3**.

After some back-and-forth between the analytical model and experimental results and tuning and calibrating the model, a good agreement between the results was finally reached. The results of the experimental model and simulation for Case No. 1 are depicted in **Figure 6**. In both models, the failure mode was due to the brace's buckling.

4. Results and discussion

The efficacy of the X-shaped concrete brace in reducing lateral displacement was investigated through simulated analysis using ABAQUS FEA modeling. The existing frame and the retrofitted frame were compared to evaluate the performance of the bracing system. The results are provided in **Figures 7–9**.

By comparing the existing frame and the retrofitted frame, the results showed that this system is suitable for reducing lateral displacement. Based on the graphs provided in **Figures 7–9**, the lateral displacement in the primary frame was 324 mm, while it was reduced by employing PPCB in Model No. 1 to 20 mm and in Model No. 2 to 17. In addition, the frame capacity to withstand lateral force was increased from 790 to 975 KN, a 23% increase, when retrofitted with concrete braces in model No. 1, while in model No. 2, it increased to 1133 KN, a 43% increase. Moreover, in model No. 2, the lateral displacement was decreased to 17 mm, which shows a significant reduction in comparison with the primary frame.

The two cases were loaded till full failure, based on the force-displacement graph provided in **Figure 8**, which shows the force-displacement of primary frames No. 1 and No. 2, and based on the displacement graph depicted in **Figure 9**, the braces were effective in reducing lateral displacements. In case No. 1, the brace is effective in recusing the lateral displacement until its failure. In model No. 2, where the compressive strength of the brace was about four times than that of model No. 1, the

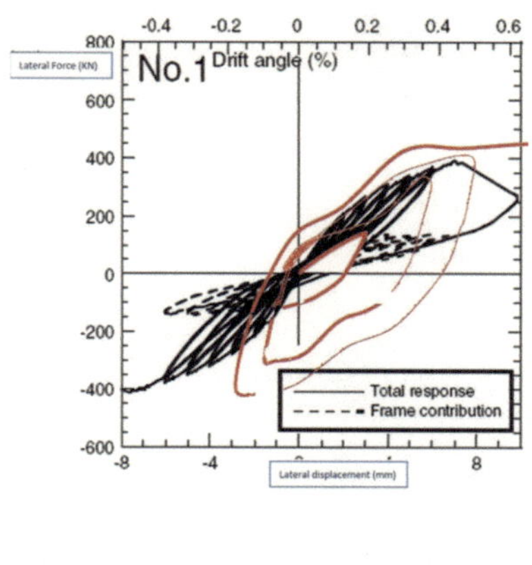

Figure 6.
Evaluating the FE modeling with experimental work; the lateral force-displacement graph of the experimental work and models No. 1 [29, 30].

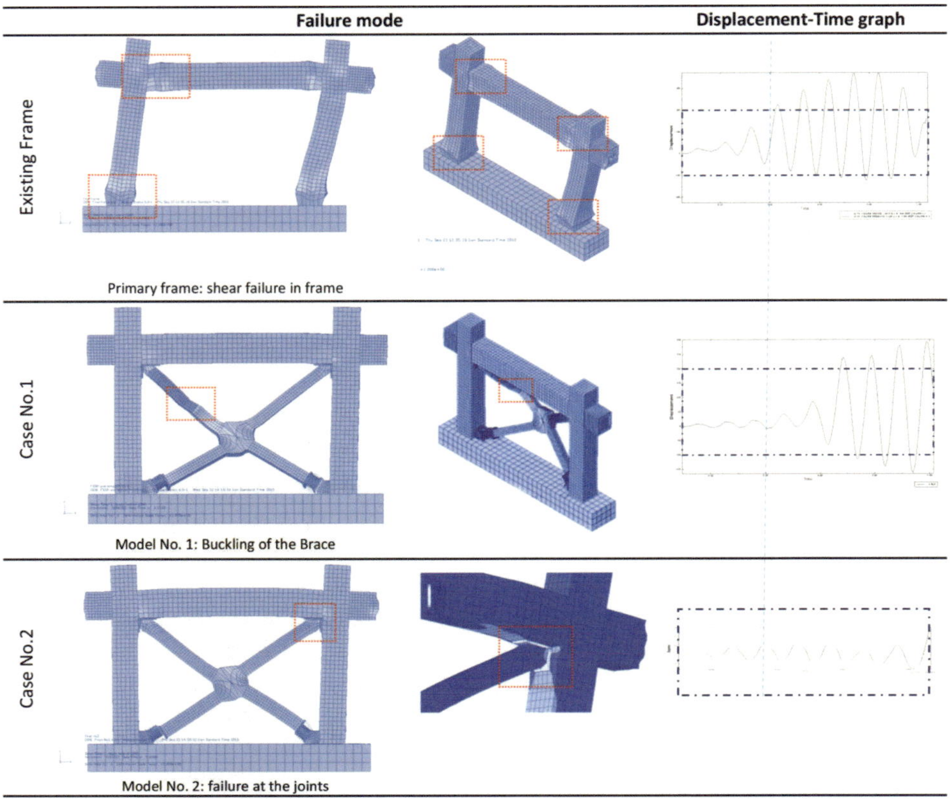

Figure 7.
Failure mode and displacement-time graph for the primary frame and models No. 1 and No. 2, modified from Refs. [29, 30].

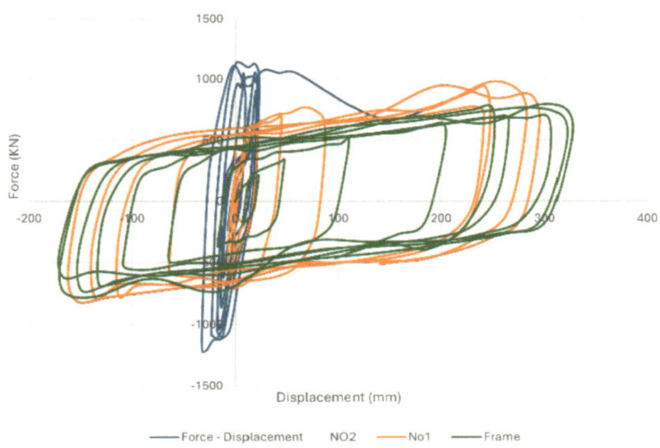

Figure 8.
The force-displacement graphs for the X-shape PPCB in the primary frame, models No. 1 and No. 2 [29, 30].

brace showed considerable stiffness, and it was more effective in lateral displacement reduction.

The test results showed that the PPCB brace effectively enhanced the initial stiffness, peak strength, and energy dissipation capacity of the original frame. The results

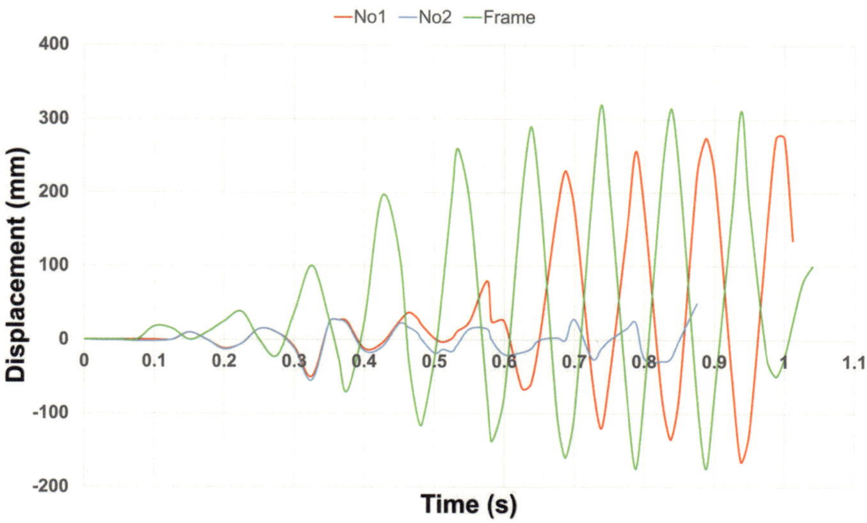

Figure 9.
The lateral displacement over time graphs for the primary frame, models No. 1 and No. 2 [29, 30].

	$f_c^{'}$ (MPa) compression strength		
Model	**Frame**	**Brace**	**Ratio (Brace/frame)**
FC 0.5	27.5	13.75	0.5
FC 0.9	27.5	24.9	0.9
FC 1	27.5	27.5	1
FC 1.5	27.5	41.25	1.5
FC 1.7	27.5	46.75	1.7
FC 2	27.5	55.0	2
FC 2.56	25.0	64.1	2.56

Table 4.
A compilation of the models used to assess the influence of brace performance-related material properties.

indicated that this technique effectively reduces lateral displacement. Therefore, the use of precast pre-stressed concrete braces is a reliable approach for retrofitting and enhancing the seismic resistance of existing concrete structures.

Further in this research, to study how the material properties such as compression strength and elastic modulus of the concrete brace will affect its' performances, we have created several models (**Table 4**) with different properties and compared the result with the primary frame. The elastic modulus for these models was calculated by Eq. (1), based on [41, 42]:

$$E_c = 5000.\sqrt{f_c^{'}} \tag{1}$$

E_c is the elastic modulus (MPa), and $f_c^{'}$ is the compressive strength of concrete (MPa).

The results indicated that in the model, the brace's compressive strength was less than that of the frame; the braces showed little stiffness and were not effective in reducing lateral displacement; and the seismic performance of the retrofitted frame was about the same as that of the primary frame, whereas in the models where the compressive strength of the brace was significantly higher than that of the primary frame, two times or more, the retrofitting method was the most effective. Additionally, the failure mode in these latest models was frame failure, in contrast to models with lesser compressive strength, which failed due to brace failure.

In the next step of this research, we have proposed and evaluated a novel configuration of the precast pre-stressed concrete braces by omitting the middle section. In this proposed single diagonal model, the middle section was removed, and instead of four legs, we will only have one member in each frame, two in a row, as depicted in **Figure 10**. This configuration has the advantage that the middle section is removed; therefore, the fabrication and installation will be a lot simpler. Similar to models No. 1 and No. 2, a FSSP device at one end and steel plates at the other end were used; columns were pre-stressed to simulate the gravity loads from the above stories; and top beams were pre-stressed to not fail early in the test. A model of this proposed bracing system was simulated in ABAQUS FEA, and the results were analyzed. In our research, two modifications to the X-shaped brace, the single diagonal and the V-shape, were proposed and evaluated.

While the frame is under lateral loading conditions, only one brace will be active in compression, while the other one, because of the FSSP device as explained before, will be elongated to fit the frame. To evaluate the proposed system, two models with different compressive strengths were created and compared to the base frame.

Based on **Figure 11**, the primary frame failed due to shear failure in column and the retrofitted frame failed due to brace buckling. A comparison of the primary frame and the single diagonal retrofitted frame shows that the single diagonal braces were effective in lateral load reduction.

While comparing it to the X-brace brace No. 1, which had the same properties and cross section, due to the increase in length of the brace, which increases the slenderness ratio, the brace bucked earlier and could not withstand lateral loads like the X-shape No. 1 brace.

We can conclude that the single diagonal concrete brace is effective in reducing lateral displacements. However, if we design the single diagonal braces in a way that buckles later than the frame, they will perform even more optimally. Therefore, in model No. 2, the compressive strength increased from 27.9 to 82.7 MPa; the results are provided in **Figure 11d**.

In this configuration, we will have only two longer parts (legs) without losing so much of their strength and a less complex installation (**Figure 8**).

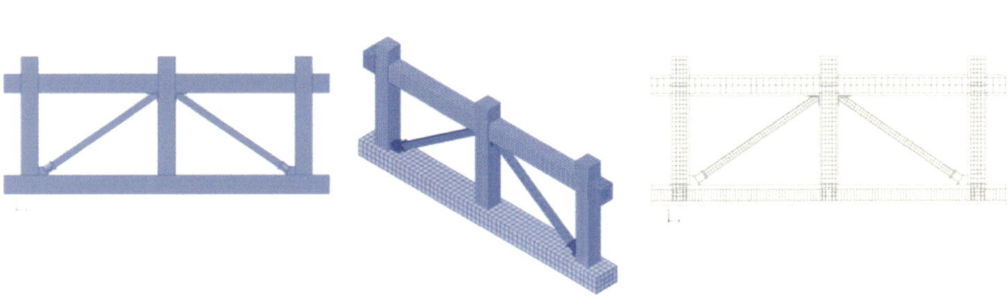

Figure 10.
The proposed novel configuration involves the use of a PPCB as a single diagonal brace by eliminating the central component and applying it to the two following frames.

In the last step of this research, to address the weakness of the single diagonal brace, another configuration, the V shape, is proposed (**Figure 12**). In comparison with the previous proposed system, single diagonal, the brace buckling might be delayed and be able to withstand more lateral forces due to its' shorter length and

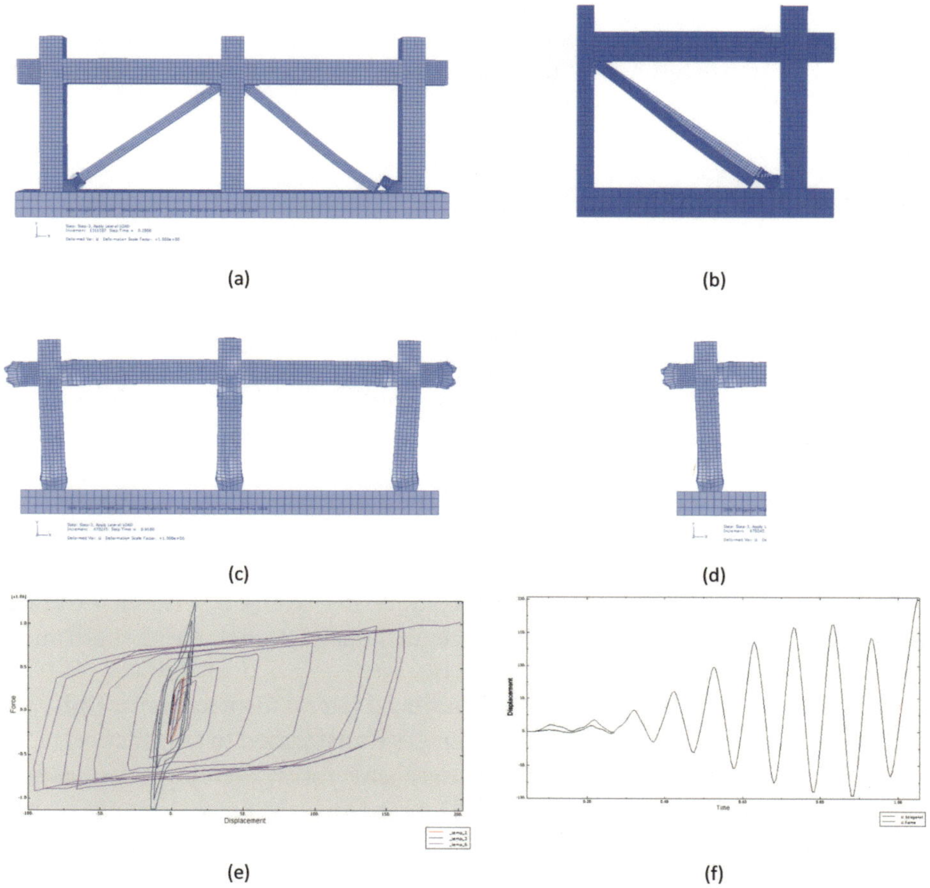

Figure 11.
Simulation results of the proposed novel configuration: single diagonal PPCB; failure mode of the strengthened frame (a), (b); failure mode of the primary frame (c), (d); force-displacement graphs of the primary and strengthened frame (e); displacement-time graph of the primary and strengthened frame (f); modified from Ref. [30].

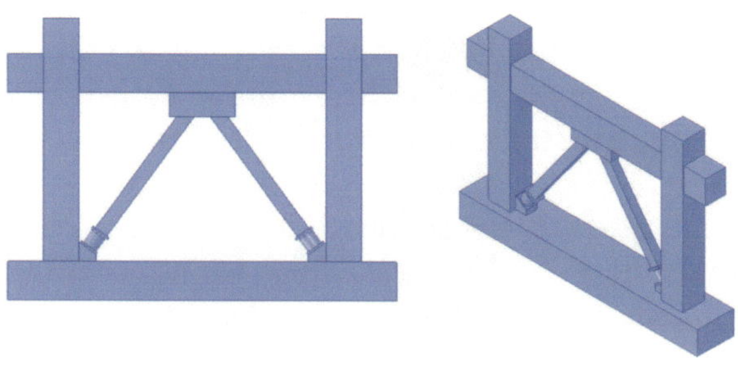

Figure 12.
Possible conceptual setup: utilizing PPCB as a brace in the shape of a V.

increased slenderness ratio for the same cross section. Moreover, the fabrication and installation of this V-shape brace in comparison with the X-shape braces are easier due to the omission of the middle section. The V-shaped concrete brace encompasses two diagonal members that are attached to the existing RC frame (beam) with a steel plate. In this configuration, the diagonal members will have a shorter length, which allows more load-bearing capacity and buckling. This faster installation will probably lead to less disturbance in usage and result in a more economical retrofitting method (**Figure 12**). The challenge here will be the concentrated upward load in the middle of the beam, which requires further studies.

5. Conclusion

Utilizing the PPCB, no wet concrete is required at the site; additionally, there is no bolt or anchorage to the existing frame; it is also simple to apply; the construction period is short; and overall costs are low. In this research project, a method for seismic retrofitting of existing buildings using precast pre-stressed concrete braces (PPCB) is *investigated and developed* by studying the seismic performance of these braces through finite element analysis using the ABAQUS software.

An analysis of the seismic performance of the brace indicated that if the compressive strength of the brace is less than that of the existing frame, the retrofitting system will have limited rigidity and will not effectively mitigate lateral displacement. However, in the scenario where the ratio of compressive strength between the brace and frame was two times or greater, the braces demonstrated notable strength and rigidity.

Results indicated that PPCB effectively reduced lateral displacement, and additionally, the force-resisting mechanism was transformed to truss action to improve its lateral stiffness and resistance to seismic force; therefore, the seismic performance of RC frames was enhanced. During the earthquake, the precast pre-stressed concrete braces can go over the plastic elongation without losing so much of their strength and can provide flexibility and capacity to dissipate lateral force energy. Thus, this method enhances stiffness and energy dissipation capacity *without disrupting buildings;* therefore, retrofitting costs could be reduced accordingly. This approach *avoids the need for wet concrete*, anchorage, or bolts while maintaining *light and ventilation,* which are often compromised by traditional methods like adding concrete shear walls. Furthermore, this method, unlike other retrofitting methods such as utilizing steel braces or FRP, which are sensitive to fire, has an improved *fire-resistance rate*. The PPCB as a retrofitting and seismic strengthening method could be *employed in various buildings' scale and usage*, including healthcare facilities, hospitals, fire stations, schools, high-rises, office buildings, and residential.

Additionally, in this research, the proposed single diagonal and V-shaped brace configuration for this system simplifies installation while increasing load-bearing capacity, resulting in a faster, more economical retrofitting process and improved fire resistance.

6. Future research

Utilizing the PPCB as a reliable method for retrofitting existing buildings is a new avenue of research and requires extensive research in the evaluation and improvement of this method. The following could be considered as the future direction for this research:

- Consider conducting additional real-world experiments to validate the FEA results.

- Explore the long-term performance and durability of PPCB under various environmental conditions.

- Using FRP on the PPCB to enhance its performance.

- Employing high-performance concretes for braces

- Detailed finite element analysis of the steel end plates.

Acknowledgements

The authors express their gratitude for the help received from the International Institute of Earthquake Engineering and Seismology in Tehran, Iran.

Author details

Mohsen Vatandoost[1,2]* and Fariborz Nateghi-A[3]

1 Taubman College of Architecture and Urban Planning, University of Michigan, Ann Arbor, USA

2 Faculty of Architecture, Department of Architecture Technology, College of Fine Arts, University of Tehran, Tehran, Iran

3 International Institute of Earthquake Engineering and Seismology, Tehran, Iran

*Address all correspondence to: mohsenv@umich.edu

References

[1] Dumaru R, Rodrigues H, Varum H. Comparative study on the seismic performance assessment of existing buildings with and without retrofit strategies. International Journal of Advanced Structural Engineering. 2018;**10**:439-464

[2] Lakshmanan N. Seismic evaluation and retrofitting of buildings and structures. ISET Journal of Earthquake Technology. 2006;**43**(1):31-48

[3] ElGawady M, Lestuzzi P, Badoux M. A review of conventional seismic retrofitting techniques for URM. In: 13th International Brick and Block Masonry Conference, Amsterdam. Vol. 4. No. 7, July 2004. Available from: https://citeseerx.ist.psu.edu/document?repid=rep1&type=pdf&doi=9ef83213c906df81a6e9d48bafdf9d7f7fcfe85f

[4] Fardis MN. Seismic design, assessment and retrofitting of concrete buildings: Based on EN-Eurocode 8. Vol. 8. Berlin: Springer; 2009. DOI: 10.1007/978-1-4020-9842-0

[5] Vatandoost M, Nateghi FA. Developing a method of strengthening concrete structures: Precast pre-stressed concrete braces. In: Presented at the 2nd National Conference on Civil Engineering, Architecture, and Urban Development of the Islamic World Countries, Tabriz, Iran, 2019. 2019. [Online]. Available from: https://civilica.com/doc/1021385/

[6] Nath S, Debnath N, Choudhury S. Methods for improving the seismic performance of structures-a review. IOP Conference Series: Materials Science and Engineering. 2018;**377**(1):012141

[7] Truong GT, Kim J-C, Choi K-K. Seismic performance of reinforced concrete columns retrofitted by various methods. Engineering Structures. 2017;**134**:217-235

[8] Cao X-Y, Feng D-C, Wu G. Seismic performance upgrade of RC frame buildings using precast bolt-connected steel-plate reinforced concrete frame-braces. Engineering Structures. 2019;**195**:382-399

[9] Gkournelos P, Triantafillou T, Bournas D. Seismic upgrading of existing reinforced concrete buildings: A state-of-the-art review. Engineering Structures. 2021;**240**:112273

[10] Navya G, Agarwal P. Seismic retrofitting of structures by steel bracings. Procedia Engineering. 2016;**144**:1364-1372

[11] Durucan C, Dicleli M. Analytical study on seismic retrofitting of reinforced concrete buildings using steel braces with shear link. Engineering Structures. 2010;**32**(10):2995-3010

[12] Basereh S, Okumus P, Aaleti S. Seismic retrofit of reinforced concrete shear walls to ensure reparability. In: Structures Congress 2020. VA: American Society of Civil Engineers Reston; 2020. pp. 498-509

[13] Al-Washali H, Suzuki Y, Maeda M. Seismic evaluation of reinforced concrete buildings with masonry infill wall. In: 16th World Conference on Earthquake, 16 WCEE 2017. Santiago Chile, January 9th to 13th 2017. Paper N° 788, Registration Code: S-N1462950003. Available from: https://www.wcee.nicee.org/wcee/article/16WCEE/WCEE2017-788.pdf

[14] Ali W, Shah SAA, Shahzada K, Ali SM, Khan SW. Retrofitting of infilled

frame in reinforced concrete structure. Mehran University Research Journal Of Engineering & Technology. 2020;**39**(3):475-488

[15] Alam Z, Zhang C, Samali B. The role of viscoelastic damping on retrofitting seismic performance of asymmetric reinforced concrete structures. Earthquake Engineering and Engineering Vibration. 2020;**19**:223-237

[16] Shakouri Mahmoudabadi N. The use of viscous dampers for retrofitting of reinforced concrete frames. Turkish Journal of Computer And Mathematics Education. 2021;**12**(13):7739-7744

[17] Noureldin M, Memon SA, Gharagoz M, Kim J. Performance-based seismic retrofit of RC structures using concentric braced frames equipped with friction dampers and disc springs. Engineering Structures. 2021;**243**:112555

[18] Marrazzo PR, Montuori R, Nastri E, Benzoni G. Advanced seismic retrofitting with high-mass-ratio tuned mass dampers. Soil Dynamics and Earthquake Engineering. 2024;**179**:108544

[19] Ferraioli M, Mandara A. Base isolation for seismic retrofitting of a multiple building structure: Design, construction, and assessment. Mathematical Problems in Engineering. 2017;**2017**(1):4645834

[20] Biddah A, Ghobarah A, Aziz TS. Upgrading of nonductile reinforced concrete frame connections. Journal of Structural Engineering. 1997;**123**(8):1001-1010

[21] Ghobarah A, Aziz TS, Biddah A. Rehabilitation of reinforced concrete frame connections using corrugated steel jacketing. Structural Journal. 1997;**94**(3):282-294

[22] Alcocer SM, Jirsa JO. Strength of reinforced concrete frame connections rehabilitated by jacketing. ACI Structural Journal. 1993;**90**(3):249-261

[23] Ozcelik R, Akpınar U, Binici B. Seismic retrofit of deficient RC structures with internal steel frames. Advances in Structural Engineering. 2011;**14**(6):1205-1222

[24] Menna C et al. Review of methods for the combined assessment of seismic resilience and energy efficiency towards sustainable retrofitting of existing European buildings. Sustainable Cities and Society. 2022;**77**:103556

[25] Comartin CD, Niewiarowski RW, Freeman SA, Turner FM. Seismic evaluation and retrofit of concrete buildings: A practical overview of the ATC 40 document. Earthquake Spectra. 2000;**16**(1):241-261

[26] Watanabe F, Miyazaki S, Tani M, Kono S. Seismic strengthening using precast prestressed concrete braces. In: 13th World Conference on Earthquake Engineering. Vancouver, B.C., Canada. August 1-6, 2004. pp. 635-642. Paper No. 3406. Available from: https://www.iitk.ac.in/nicee/wcee/article/13_3406.pdf

[27] Elnashai AS, Di Sarno L. Fundamentals of Earthquake Engineering: From Source to Fragility. Printed in England by Antony Rowe Ltd, Chippenham, Wilts. John Wiley & Sons; 2015. ISBN: 978-0-470-02483-6 (Hbk). Available from: https://thuvienxaydung.net/files/document/kdoan/fundamentals_of_earthquake_engineering_7a387B_09esk9g7.pdf

[28] Sarfarazi V, Ghazvinian A, Schubert W, Nejati HR, Hadei R. A new approach for measurement of tensile strength of concrete. Periodica

Polytechnica Civil Engineering. 2016;**60**(2):199-203

[29] Nateghi FA, Vatandoost M. Seismic retrofitting RC structures with precast prestressed concrete braces-ABAQUS FEA modeling. International Journal of Engineering. 2018;**31**(3):394-404

[30] Nateghi Alahi F, Vatandoost M. Single diagonal precast prestressed concrete bracing for strengthening existing concrete frames. International Journal of Advanced Structural Engineering. 2018;**10**(4):339-347

[31] Vatandoost M. Seismic retrofitting RC structures with precast prestressed concrete braced [thesis]. Iran: Azad University, Takestan Branch; 2010

[32] Vishnuprakash N, Ramesh T. study on the behaviour of various types of prestressed flat slab under single storey frame with and without braces under seismic effects. In: Fonseca de Oliveira Correia JA, Choudhury S, Dutta S. editors. Advances in Structural Mechanics and Applications. ASMA 2021. Structural Integrity. Vol. 26. Cham: Springer; 2023. DOI: 10.1007/978-3-031-05509-6_29

[33] Borah M, Choudhury S. Seismic performance of tall buildings with different structural systems. In: International Conference on Advances in Structural Mechanics and Applications. Cham: Springer International Publishing; Mar 2021. pp. 90-107. DOI: 10.1007/978-3-031-05509-6_8

[34] Lee B-G, Kim J-Y, Jung J-S, Lee K-S. Seismic protection provided by a new diamond-shaped bracing system with a horizontally layered friction damper. Journal of Building Engineering. 2024;**92**:109709

[35] Zhang Y, Hu Y, Li N, Xie L, Wang Z, Liu D. Isolation performance evaluation of base-isolated system with active nonlinear negative stiffness devices. Soil Dynamics and Earthquake Engineering. 2024;**179**:108565

[36] Chong X, Liu H, Xie L, Zhou A, Huang J, Sha H. Experimental investigation of retrofitting seismically damaged reinforced-concrete frames using steel jacketing. Structure. 2024;**62**:106243: Elsevier

[37] Lee B-G, Kim J-Y, Jung J-S, Lee K-S. Seismic capacity of R/C buildings retrofitted with a V-bracing system equipped with a novel laterally layered friction damper. Applied Sciences. 2023;**13**(24):13205

[38] Maheri MR, Najafgholipour M, Zarandi S. Seismic rehabilitation of URM heritage-listed Namazi school building using multiple retrofitting techniques. Journal of Building Engineering. 2023;**79**:107782

[39] Hung C-C, Hsiao H-J, Shao Y, Yen C-H. A comparative study on the seismic performance of RC beam-column joints retrofitted by ECC, FRP, and concrete jacketing methods. Journal of Building Engineering. 2023;**64**:105691

[40] Cao X-Y, Feng D-C, Wu G, Wang Z. Experimental and theoretical investigations of the existing reinforced concrete frames retrofitted with the novel external SC-PBSPC BRBF sub-structures. Engineering Structures. 2022;**256**:113982

[41] Vakhshouri B. Modulus of elasticity of concrete in design codes and empirical models: Analytical study. Practice Periodical on Structural Design and Construction. 2018;**23**(4):04018022

[42] Architectural Institute of Japan. Standards for structural calculation of steel reinforced concrete structures. 1987

Chapter 8

Surface Layer Concrete

Noboru Yuasa

Abstract

Regarding the quality of structural concrete that is demolded during hydration, the porosity is not as dense as the surface layer because hydration is inhibited by drying. As a result, the strength of the surface layer is lower than that of the inside, and external deterioration factors easily enter, resulting in a decrease in deterioration resistance. In addition, since the surface concrete serves as the substrate for finishing materials, it is associated with problems such as adhesion and blistering of the finish. This chapter describes the inhomogeneity of moisture content and pore structure of concrete from the surface to the interior, strength distribution, resistance to deterioration, and peeling resistance of finishing materials.

Keywords: concrete, surface, inhomogeneity, moisture content, porosity, strength, deterioration, isothermal moisture adsorption and desorption, permeability, peeling of finishing

1. Introduction

By practical reasons, formworks in concrete constructions are removed at early ages, when hydration reactions are still proceeding. Because drying initiates from the surface of the concrete, it results shortage of water necessary for hydration at a position near the surface. As a result, the surface part of concrete is likely to have less moisture content and coarser pore structure than those of the interior parts. These coupled phenomena may affect strength, evolution, and durability, and are closely related to the delamination-separation of finishing.

Though the effects of drying at early ages manifest as a function of depth from the surface, past studies [1] only show relationships between curing conditions and resulting strength or durability as a whole, and the difference of moisture conditions and pore structure as functions of depth from the surface have never been extensively investigated.

This chapter deals with the effect of the drying initiation age on the inhomogeneity of concrete from the standpoint of moisture content and pore structure, and discusses the effect of the quality of the surface concrete on strength distribution, deterioration of the structure, and peeling of finishing materials.

2. Age of drying initiation and inhomogeneity

The inhomogeneity of concrete subjected to drying is studied in a wide variety of mixture proportions and drying conditions [2]. The water-cement ratio is 0.3, 0.4, 0.6, and 0.8, and the age of drying initiation is 1, 3, 7, and 28 days after mixing. The resulting variations of moisture content and pore structure were determined during 1 day to 1 year after mixing, respectively.

Ordinary Portland cement, river sand, river gravel, and chemical admixture are mixed at different water-cement ratios. After casting concrete, specimens were set in an air-conditioned room at a temperature of 20°C and relative humidity of 60%. Each end-side, 10 × 10 cm, of specimens was opened to the indoor air for drying at the age of 1, 3, and 7 days.

Moisture content of the specimens was determined non-destructively by the ceramic moisture sensors as shown in **Figure 1** [3]. Bound water, the non-evaporable water incorporated in the hydration products, were assumed to be ignition loss at 600°C. The bound water content in unit mass of hardened cement paste was defined as a measure of degree of hydration. Specimens of 1 cm thick were cut out at a depth of 0–1, 2–3, 4.5–5.5, 9.5–10.5, 19.5–20.5 from the drying surface, adjusted to a particle size between 2.5 and 5.0 mm, treated by acetone and then D-dried.

The determination procedure of bound water content was to weigh the mass of a specimen before, W_0 (g), and after, Wi (g), heating to 600°C for 1 hour. The heated specimens were then mixed with 10% hydrochloric acid and stirred for 2 hours to dissolve cement components [4].

Subsequently, the solution was heated at 600°C for 1 hour, cooled in a desiccator, and weighed as an insoluble residue Wns (g). The soluble component content (cement paste content in g/g) WRs and the bound water content WRh (g/g) can be calculated by Eqs. (1) and (2), respectively.

$$\mathrm{WRs} = (W_0 - \mathrm{Wns})/W_0 \tag{1}$$

$$\mathrm{WRh} = (W_0 - \mathrm{Wi})/W_0 \times \mathrm{WRs} \tag{2}$$

Specimens were subjected to mercury porosimetry to determine pore structure ranging from 3 nm to 3.2 × 10^5 nm in radius. Effective pore volume Vep (cc/g) that is the pore volume present in the cement paste in the specimen can be calculated from the measured pore volume Vmp (cc/g) with Eq. (3) [4].

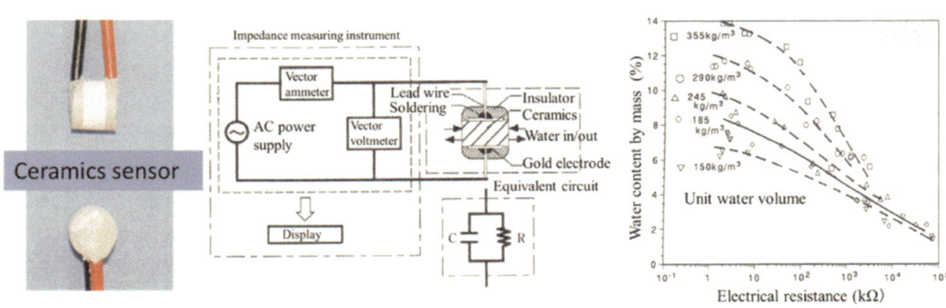

Figure 1.
Embedded ceramics sensor.

$$\text{Vep} = \text{Vm}/\text{WRs} \tag{3}$$

2.1 Distribution of moisture content

Changes in moisture content with time are shown in **Figure 2** with respect to drying initiation ages and water-cement ratio. All specimens show a rapid decrease in moisture content near the dying surface immediately after the initiation of drying. At the age of 28 days, the extent of the decrease in moisture content is approximately within 5 cm from the surface without regard to water-cement ratio. These tendencies are consistent with the moisture content distribution calculated and analyzed by Sakata et al. [5]. Considering the hydration rate becomes slow after 28 days, the hindered hydration zone can be estimated as 5 cm from the drying surface. When the water-cement ratio is as high as 0.6 or 0.8 and the drying initiation age is as early as 1 day, moisture content of concrete more than 5 cm from the drying surface changes gently according to the distance from the surface.

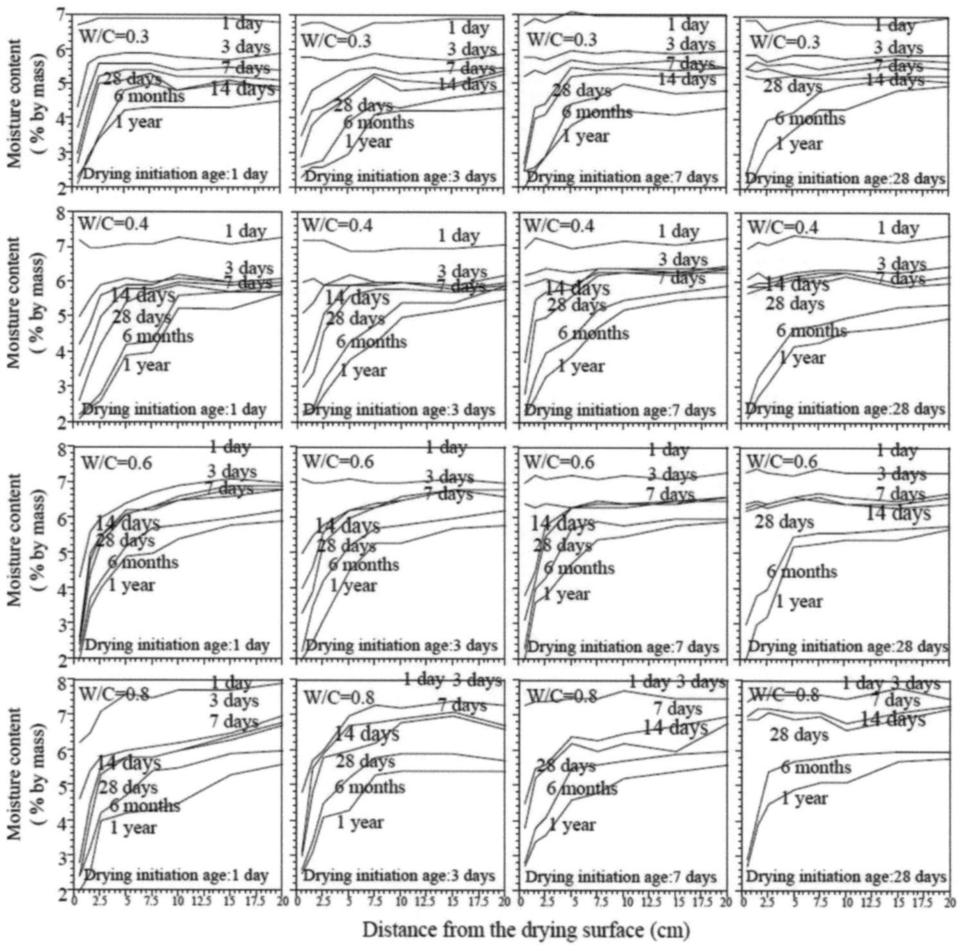

Figure 2.
Effects of water-cement ratio and drying initiation age on moisture content.

Long-term moisture distribution, with little effects of drying initiation ages, resulted in an equilibrium moisture content which is a function of ambient temperature and relative humidity.

Specimens kept in a constant hygro-thermal condition, temperature of 20°C and relative humidity of 60% for 1 year, resulted in a moisture content of 2.2% at 0.5 cm from the drying surface without regard to water-cement ratio and drying initiation ages. However, moisture content at the center of the specimen, 20 cm from the drying surface, shows a larger value according to the water-cement ratio. It is 4.6 and 5.5% when water-cement ratio was 0.3 and 0.8, respectively.

Even before drying initiation, decrease in moisture content uniformly from the surface to the central part is observed, which is more conspicuous at a low water-cement ratio and can be attributed to the self-desiccation. Self-desiccation is an apparent drying phenomenon caused by the consumption of evaporation of water by hydration. Therefore, the moisture content measured in this way is the consequence of evaporable water changes both by drying and hydration.

2.2 Distribution of bound water

Distribution of bound water at the age of 28 days is shown in **Table 1**. When the drying initiation age is earlier and the position is closer to the drying surface, bound water content becomes smaller. Taking an undried specimen that is sealed until the age of 28 days as a reference, the bound water content of specimen dried from the age of 1 day with water-cement ratio of 0.3 is 70% of that of the reference, and that with water-cement ratio of 0.4 is 80%. When water-cement ratio is 0.8, the effect of drying initiation ages on the bound water content is within the range of scatter of the test and no particular difference is observed.

Effect of drying on the decrease of bound water is remarkable when the water-cement ratio becomes smaller.

2.3 Inhomogeneity of pore structure

The total effective pore volume of all specimens is also shown in **Table 1. Figure 3** shows the depth distribution of the total effective pore volume of concrete. It can be seen that the level differs when the water-cement ratio is different and that the surface layer of dried concrete has a coarse structure. This is more pronounced the earlier drying begins. The surface layer of 1 cm on the first day of drying corresponds to an internal pore volume that is 0.2 greater in terms of water-cement ratio.

The range where the total effective pore volume is high is within a depth of 5 cm in typical concrete with a water-cement ratio of 0.6 or less, but in poor quality concrete, it extends to deeper layers.

Pore size distributions at the age of 28 days are shown in **Figure 4**. Samples near the drying surface show pore size distribution shifting to a larger pore side.

The median pore radius in **Table 1** is defined as a pore radius corresponding to 50% of pore volume accumulated from the smaller side representing the deviation of distribution. It ranged from 20 nm to 30 nm for specimens without drying and was slightly affected by the water-cement ratio, but when subjected to drying, the effect of the drying initiation ages and the distance from the drying surface are significant resulting in larger median pore radius as the drying effect became larger. The median pore radius shift was greatly dependent on water-cement ratio.

W/C	Drying initiation age (day)	distance from the surface (cm)	Bound water content (g/g)	Vep × 10^{-4} (cc/g)	Median pore radius (nm)	W/C	Drying initiation age (day)	distance from the surface (cm)	Bound water content (g/g)	Vep × 10^{-4} (cc/g)	Median pore radius (nm)
0.3	1	0–1	0.154	1599	38.6	0.4	1	0–1	0.162	2208	63.7
		2–3	0.180	1272	29.9			2–3	0.187	1966	35.0
		4.5–5.5	0.201	1241	25.1			4.5–5.5	0.215	1674	23.8
		9.5–10.5	0.196	1180	23.6			9.5–10.5	0.214	1645	23.9
		19.5–20.5	0.190	1128	23.7			19.5–20.5	0.212	1576	23.5
	3	0–1	0.159	1400	36.4		1	0–1	0.172	2062	71.5
		2–3	0.183	1211	27.8			2–3	0.200	1716	38.5
		4.5–5.5	0.180	1177	24.7			4.5–5.5	0.217	1673	23.3
		9.5–10.5	0.178	1172	26.6			9.5–10.5	0.206	1609	22.8
		19.5–20.5	0.182	1151	22.8			19.5–20.5	0.206	1616	24.4
	7	0–1	0.171	1309	42.3		7	0–1	0.196	1744	44.4
		2–3	0.175	1297	25.7			2–3	0.215	1707	29.1
		4.5–5.5	0.184	1261	24.8			4.5–5.5	0.220	1688	23.0
		9.5–10.5	0.183	1206	29.8			9.5–10.5	0.223	1659	24.0
		19.5–20.5	0.179	1177	24.5			19.5–20.5	0.213	1568	25.0
	Sealed	0–1	0.184	1290	24.4		Sealed	0–1	0.196	1531	32.8
		2–3	0.188	1210	24.0			2–3	0.213	1644	27.6
		4.5–5.5	0.183	1183	24.3			4.5–5.5	0.214	1620	23.3
		9.5–10.5	0.188	1285	26.6			9.5–10.5	0.211	1645	20.0
		19.5–20.5	0.187	1191	23.8			19.5–20.5	0.209	1627	21.7

W/C	Drying initiation age (day)	distance from the surface (cm)	Bound water content (g/g)	Vep × 10^{-4} (cc/g)	Median pore radius (nm)	W/C	Drying initiation age (day)	distance from the surface (cm)	Bound water content (g/g)	Vep × 10^{-4} (cc/g)	Median pore radius (nm)
0.6	1	0–1	0.194	3644	163.1	0.8	1	0–1	0.204	5053	265.4
		2–3	0.211	2651	153.5			2–3	0.197	4898	153.6
		4.5–5.5	0.212	2541	56.1			4.5–5.5	0.211	4646	56.2
		9.5–10.5	0.215	2624	30.7			9.5–10.5	0.205	3691	35.0
		19.5–20.5	0.214	2631	35.8			19.5–20.5	0.197	3609	35.8
	3	0–1	0.186	3340	174.0		3	0–1	0.211	4435	174.0
		2–3	0.206	2646	86.7			2–3	0.199	4256	86.7
		4.5–5.5	0.207	2557	54.3			4.5–5.5	0.207	4214	37.2
		9.5–10.5	0.208	2588	31.7			9.5–10.5	0.199	3699	31.7
		19.5–20.5	0.208	2669	22.9			19.5–20.5	0.215	3661	32.9
	7	0–1	0.185	2669	108.6		7	0–1	0.193	4055	108.6
		2–3	0.204	2754	54.3			2–3	0.204	3880	61.9
		4.5–5.5	0.207	2576	32.1			4.5–5.5	0.202	3794	32.1
		9.5–10.5	0.205	2575	27.5			9.5–10.5	0.204	3666	275.
		19.5–20.5	0.204	2540	27.2			19.5–20.5	0.190	3654	27.2
	Sealed	0–1	0.201	2691	20.3		Sealed	0–1	0.193	3751	28.6
		2–3	0.203	2580	22.1			2–3	0.193	3779	28.2
		4.5–5.5	0.205	2840	24.4			4.5–5.5	0.196	3779	25.1
		9.5–10.5	0.198	2653	21.7			9.5–10.5	0.197	3682	29.0
		19.5–20.5	0.204	2653	22.9			19.5–20.5	0.201	3655	28.6

Table 1.
Bound water content, total effective pore volume, and median diameter.

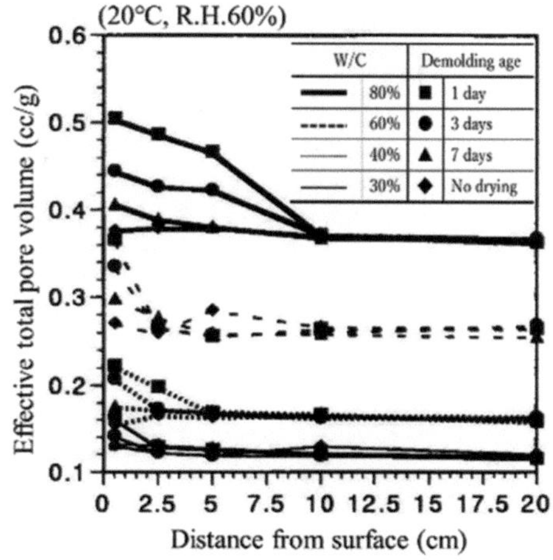

Figure 3.
Distribution of effective total pore volume (concrete age: 28 days).

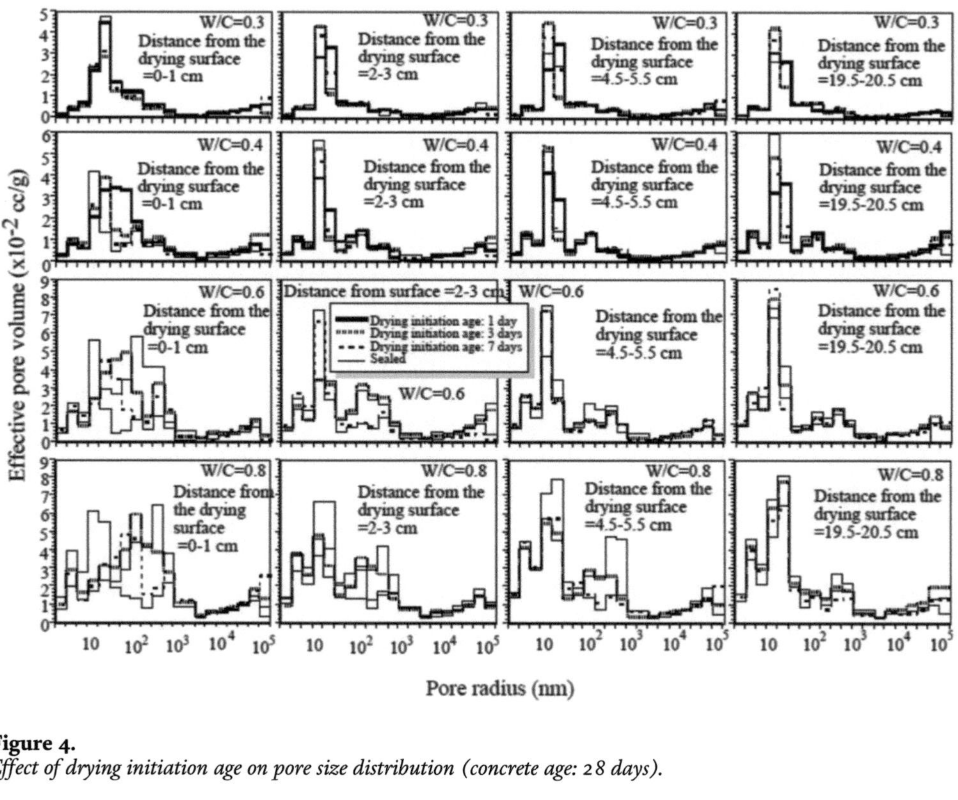

Figure 4.
Effect of drying initiation age on pore size distribution (concrete age: 28 days).

Figure 5 shows the relative effective pore volume distribution of dried concrete and non-dried concrete. The effective pore volume of dried concrete at a radius of more than 18 nm or 32 nm increases while that of the other part decreases. This may be attributed to the interference of hydration due to drying resulting in the incomplete consolidation of pores at a radius more than 18 nm–32 nm. This tendency is significant when the drying initiation age is earlier, the position is closer to the drying surface, and the increasing pore volume extends over a larger pore radius. When the

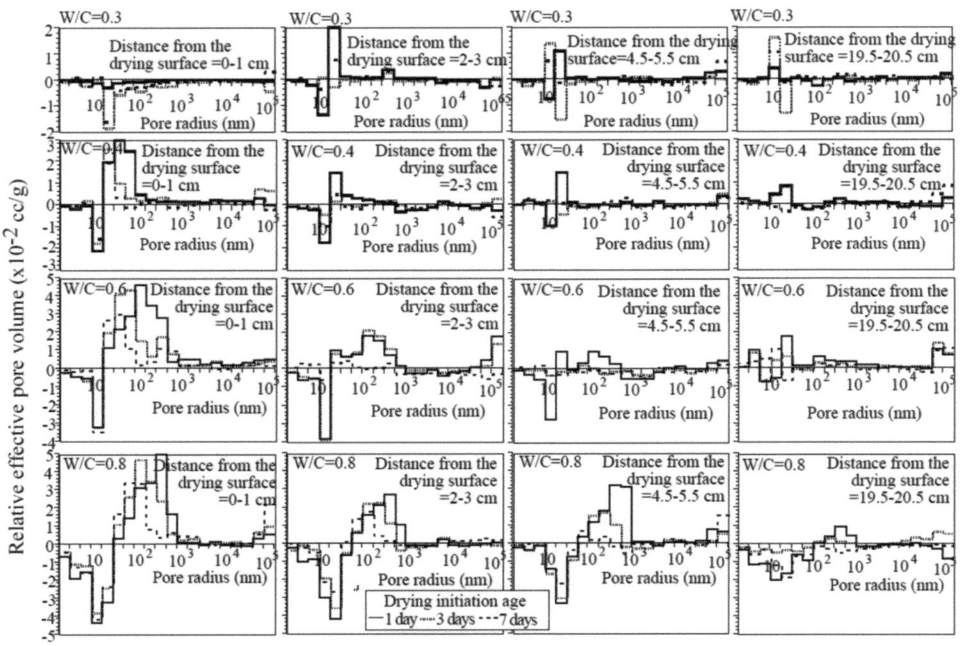

Figure 5.
Increased or decreased pore volume due to drying (20°C, R.H.60%).

water-cement ratio is large, this tendency is more conspicuous and reaches deeper portion from the drying surface. In order not to have coarser pores, the curing capable of preventing concrete at early ages from drying is necessary, especially at a larger water-cement ratio.

As shown in **Figure 6**, as concrete age passes, the internal pores become denser, but the pore distribution in the surface layer remains from when drying began.

Figure 7 shows the difference in pore distribution of surface concrete depending on the age at which drying starts, depending on the type of cement used. When retarding type cement is used, the pores remain coarser due to drying at an early age.

Although the effect of curing on the pore structure was not resolved by depth, the results of Chen et al. support the results aforesaid [6]. Chen et al. have attempted to model the pore structure based on the research results of Shimomura, Maekawa, and others [7].

3. Inhomogeneity of compressive strength

Herbert J. Gilkey's experimental results on the effect of dry and wet curing on the strength of concrete are well known, but this test was conducted on a 2 × 4 in. (5.1 × 10.2 cm mortar (70% W/C)) [1].

Many researchers have attempted to estimate the compressive strength of concrete from its pore structure [8]. **Figure 8** shows the relationship between the compressive strength of a sealed and dried 7.5 × 11.5 mm micro-cement paste specimen and the estimated compressive strength from its pore structure using Yoshino and Kamada al.'s Eq. (4) [4, 9]. It can be confirmed that it is generally possible to estimate the

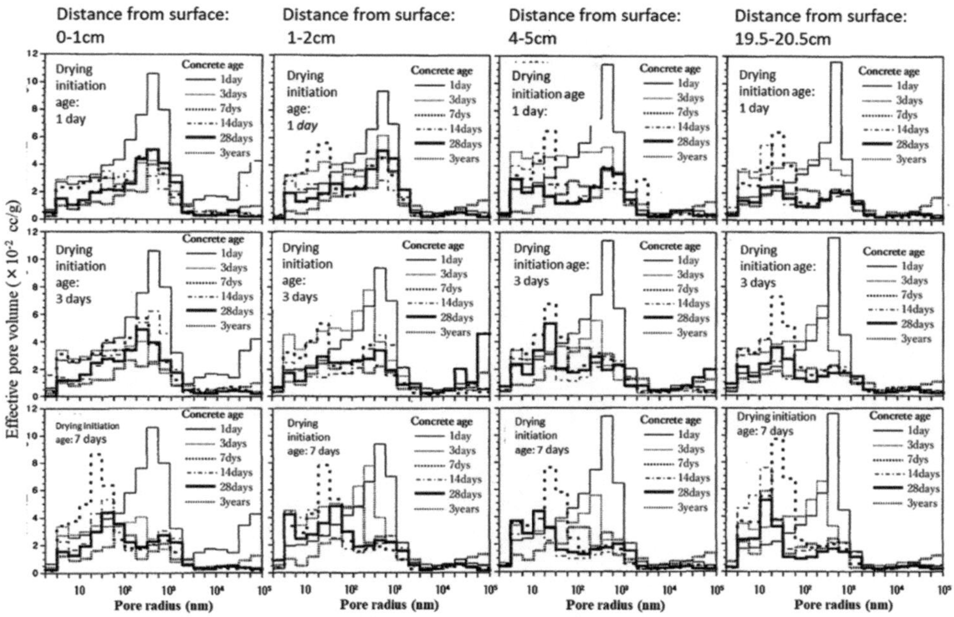

Figure 6.
Changes in the pore structure of dried concrete over time (20°C, R.H.60%).

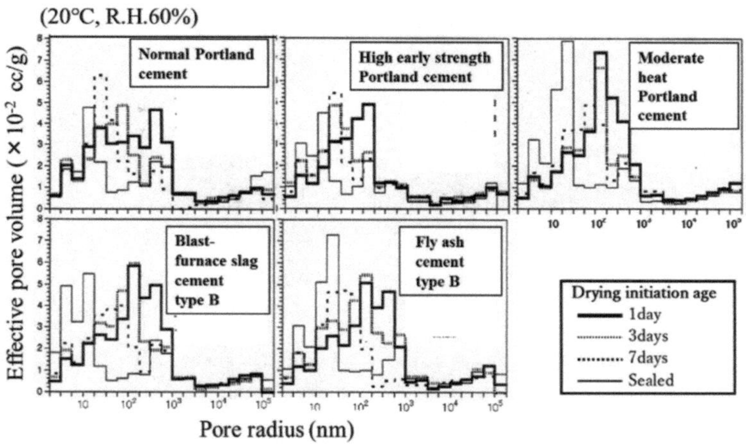

Figure 7.
Effect of cement used on pore distribution of dried surface layer concrete (0–1 cm, W/C: 60%, concrete age: 28 days).

compressive strength of hardened cement specimens with unique pore structures without enough moist curing.

Figure 9 shows the compressive strength distribution estimated by Yoshino and Kamada al.'s Eq. (4) using the data shown in **Table 1** for dried concrete [10].

$$\sigma = 144 \times \exp\left(-2.67 \times \mathrm{ETPV} - 0.485 \times \log\left(\mathrm{Me} \times 10\right) - 0.96 \times \mathrm{Re} + 3.56 \times \mathrm{WRh}\right) \quad (4)$$

where σ: Compressive strength (MPa, N/mm^2).

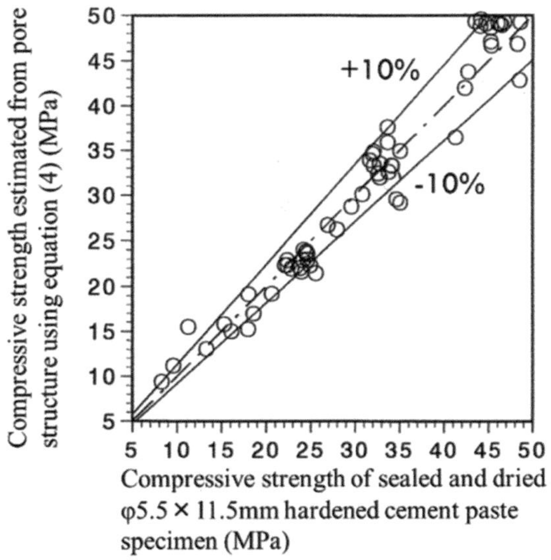

Figure 8.
Compressive strength of φ5.5 × 11.5 mm specimen sealed and dried before testing and compressive strength estimated from pore structure.

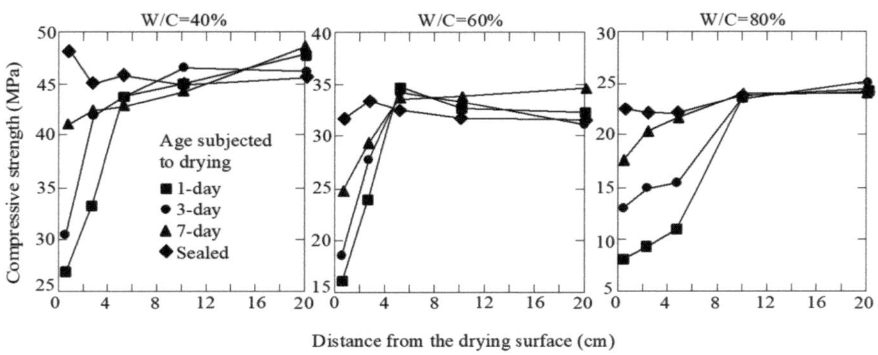

Figure 9.
Distribution of compressive strength as a function of the distance from the drying surface (20°C, R.H.60%, concrete age: 28 days).

ETPV: Effective total pore volume (cc/g).

Me: Median pore radius: Pore radius corresponding to 50% of the pore volume accumulated from the smallest side (nm).

Re: Returned ratio: Ratio of pore volume measured when pressure is reduced to pore volume measured when pressure is increased.

WRh: Bound water content (g/g).

For all strength levels of concrete, the closer to the dry surface the lower the compressive strength, but the value remains constant inside regardless of dryness. For ordinary-level concrete, a decrease in strength is observed within a range of roughly 5 cm. Furthermore, regardless of strength level, the amount of decrease is large, reaching approximately 15 N/mm^2 compared to the internal strength, which can be considered potential strength.

4. Surface layer concrete for the durability of reinforced concrete structures

Deterioration of reinforced concrete structures, as shown in **Figure 10**, occurs when deterioration factors such as carbon dioxide (stage 1) and water and oxygen (stage 2) in rebar corrosion due to neutralization, chloride ions (stage 1), water and oxygen (stage 2) in salt damage, and water in frost damage and alkali-silica reaction (stage 2) penetrate from the surface and gradually move inward. Water is essential for the hydration and hardening of cement, but as shown in **Table 2**, it becomes the root of all evil for reinforced concrete structures after the concrete has hardened [11].

Figure 10.
Deteriorations of reinforced structure.

Deterioration	Role of water	Degree of deterioration
Rebar corrosion due to carbonation	Corrosion proceeds in the presence of **water** and oxygen.	No damages without **water**.
Rebar corrosion due to chloride ion	Corrosion proceeds in the presence of excess amount of chloride ions, **water,** and oxygen.	
Frost expansion of concrete	Concrete pore pressure induced by freezing of **water**.	Degree of deterioration reduces at low **water** content.
Swelling of concrete due to ASR	Swelling of reaction products in the reactive aggregate due to osmotic pressure of free **water**	

Table 2.
Role of water and damage degree in the deterioration mechanisms.

Considering these factors, the deterioration resistance of reinforced concrete structures depends on the permeation resistance of the surface concrete to deterioration factors and its physical and chemical stability.

However, as mentioned above, compared to the interior, which can be said to have potential quality mainly due to the water-cement ratio, the surface concrete has a coarse pore structure. Therefore, in the surface layer of concrete, external deterioration factors such as carbon dioxide, chloride ions, water, and oxygen can easily move. It is assumed that the surface layer concrete also has low resistance to mechanical deterioration.

4.1 Resistance to freezing and thawing

Looking at the pore structure macroscopically in terms of the total pore volume, as shown in **Figure 3**, the total effective pore volume is constant internally depending on the water-cement ratio, but the pore volume in the surface 5 cm is large.

Figure 11 shows the scaling contents of dried concrete tested by freeze-thaw test method A (frozen at −18°C) using 10 × 10 × 40 cm specimens [12]. The scaling contents increase accordingly as concrete is dried earlier. Though reducing the

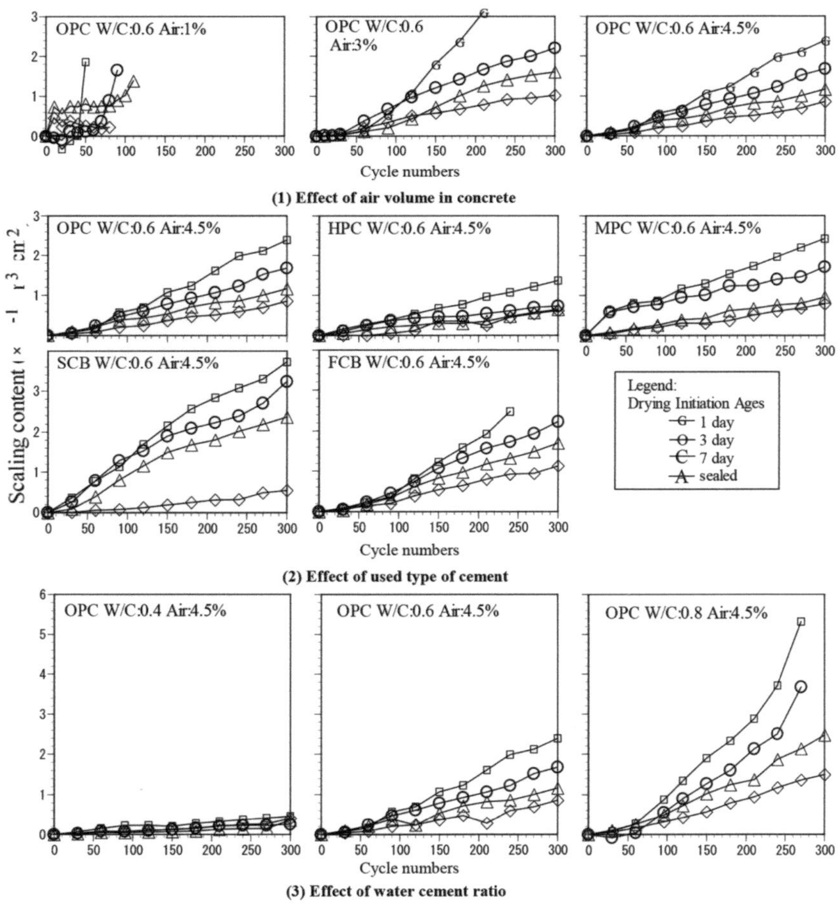

Figure 11.
Change of scaling content of dried concrete (20°C, R.H.60%).

water-cement ratio is effective in avoiding scaling, keeping moist curing is more important at early ages. In other words, after controlling the air volume and water-cement ratio to avoid expansion, keeping moist curing at early ages is more effective in avoiding scaling than reducing water-cement ratio immoderately.

In the case of using blast-furnace cement type B, the dried concretes have a remarkably higher amount of scaling than non-dried concrete. For non-dried concrete, concretes using fly ash cement type B have a higher amount of scaling. In reality, for the case of using blast-furnace cement type B and fly ash cement type B, reducing water-cement ratio sharply may be needed to resist scaling.

As air volume increases, the amount of scaling reduces. But the difference between 3 and 4.5% of air volume is small and may be improved by narrowly lowering the water-cement ratio or lengthening the moist curing term. Though keeping air volume in concrete is important to avoid expansion, it is not very effective to resist scaling.

It is easier for water in larger pores to freeze than for water in smaller pores to freeze [13]. So, the correlations between the surface layer (0–1 cm) effective pore volume in the range of larger than certain pore sizes and the scaling content per cycle were calculated.

As a result, the effective pore volume of more than 56 nm in radius is best related to the scaling as shown in **Figure 12**. This 56 nm as the lower limit radius can be considered as reasonable by the analogy from previous research studies [14, 15]. As mentioned above, the pore of larger than 56 nm in radius is mostly found in concrete dried at early ages. Accordingly, moist curing at an early age is extremely important to prevent scaling by the freezing and thawing reaction.

Changes to the relative dynamic modulus of elasticity are shown in **Figure 13**. This clearly shows that the air volume in concrete is the most important factor for freeze-thaw resistance. In the case of 3% air concrete, the relative dynamic modulus of elasticity decreases earlier when the start of drying is earlier. Though concrete involved 4.5% air, the decrease of relative dynamic modulus of elasticity of the specimen using type B Portland blast-furnace slag cement and type B Portland fly ash can be seen at the drying age of 1 and 3 days.

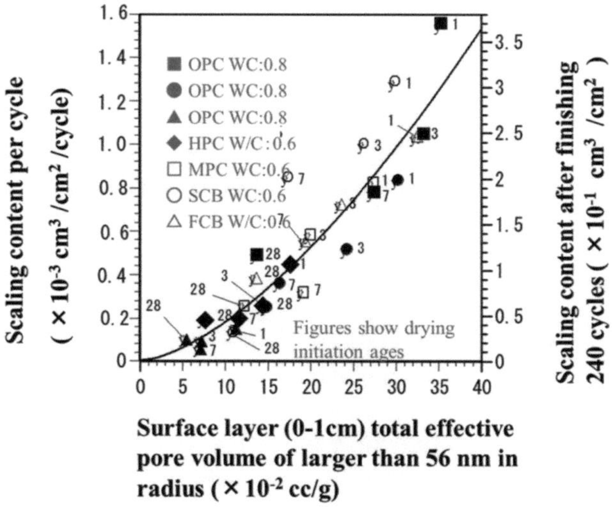

Figure 12.
Relationship between surface layer total effective pore volume of larger than 56 nm in radius and scaling content.

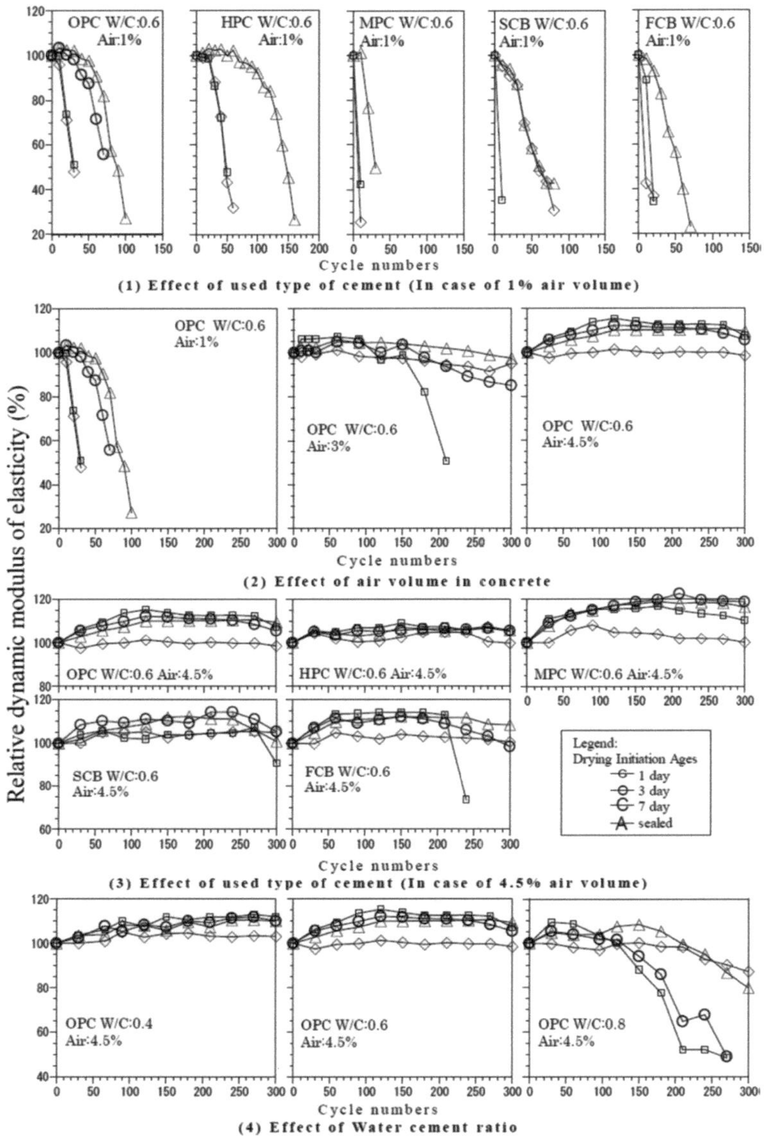

Figure 13.
Changes of relative dynamic modulus of elasticity of dried concrete (20°C, R.H.60%).

From these results, to avoid deterioration with expansion, it is important firstly to keep the 4.5% air volume and secondly to reduce the water-cement ratio.

However, by merely maintaining the air volume or reducing the water-cement ratio is not enough, early moist curing plays an important role for the resistance to freezing and thawing. Therefore, when using medium-heat Portland cement, blast-furnace slag cement, or fly ash cement, which have slow hydration times, it is important to ensure wet curing, especially at the early concrete age.

In concretes that involve 4.5% of air, the pore size distributions of seven concretes that show reducing dynamic modulus of elasticity and increasing expansion for test of freezing and thawing are shown in **Figure 14(1)** and the pore size distributions of 21

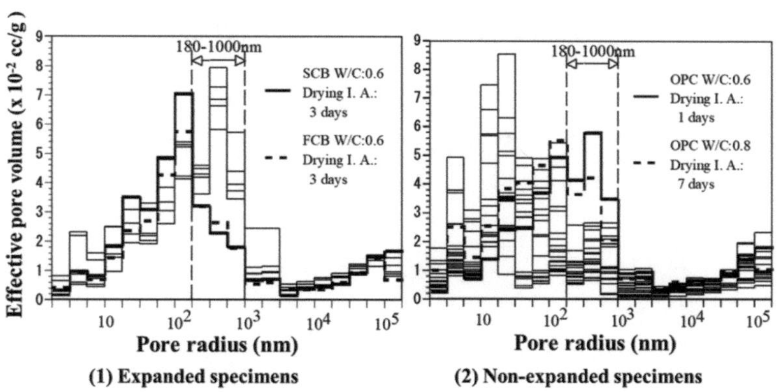

Figure 14.
Effect of pore distribution on expansion deterioration.

concretes that did not show these changes for test are shown in **Figure 14(2)**. In **Figure 14**, only surface layer (0–1 cm) distributions are shown.

Most concretes that have reduced dynamic modulus of elasticity and increased expansion have more pores in the range of 180 nm to 1000 nm in radius than concretes that do not. The respective two concretes shown by broken lines in **Figure 13(1)** and **(2)** are exceptions from this tendency. Kamada noted that the pores in the range of 75 nm to 7500 nm in radius have relationship with the frost damage [16]. For this pore, Trinker has pointed out from the range of 100 nm to 1000 nm in radius [17], Kokufu from 24 nm to 7500 nm in radius [18], and Sumiyoshi from 42 nm to 7500 nm in radius [19]. The results from this experiment roughly agree with those from other studies. Furthermore, Kamada explained this relationship from the theory of the falling of melting pot and developed it into the theory on the mechanism of frost damage [14, 15].

As mentioned above, pores of larger than 18 nm in radius, especially from 100 nm to 1000 nm, tend to increase when concrete is dried from an early age. So, to avoid expansion due to the action of freezing and thawing, wet curing of concrete at an early age is important.

4.2 Neutralization resistance

Figure 15 shows the progress of the neutralization of dried concrete. The water-cement ratio has a greater effect than the drying initiation age, and the smaller the water-cement ratio, the slower the neutralization progresses. However, for the same water-cement ratio, the earlier the drying initiation age, the faster the neutralization progresses. With a water-cement ratio of 0.3 or 0.4, neutralization does not progress at all unless the concrete is dried until 3 days after casting.

Also, looking at the effect of the cement used, the effect of the cement used is greater than the drying initiation age. Neutralization progresses slowly with high early-strength Portland cement, which hydrates quickly, while neutralization is faster with blast-furnace cement type B, fly ash cement type B, and moderate heat Portland cement, which hydrates slowly and has low alkalinity.

The neutralization index NF is defined by the Eq. (5), where P_{10} is the accelerated period (weeks) when the neutralization depth reaches 10 mm, Dc is the neutralization depth of 10 mm or the neutralization depth at 26 (mm) weeks of the accelerated

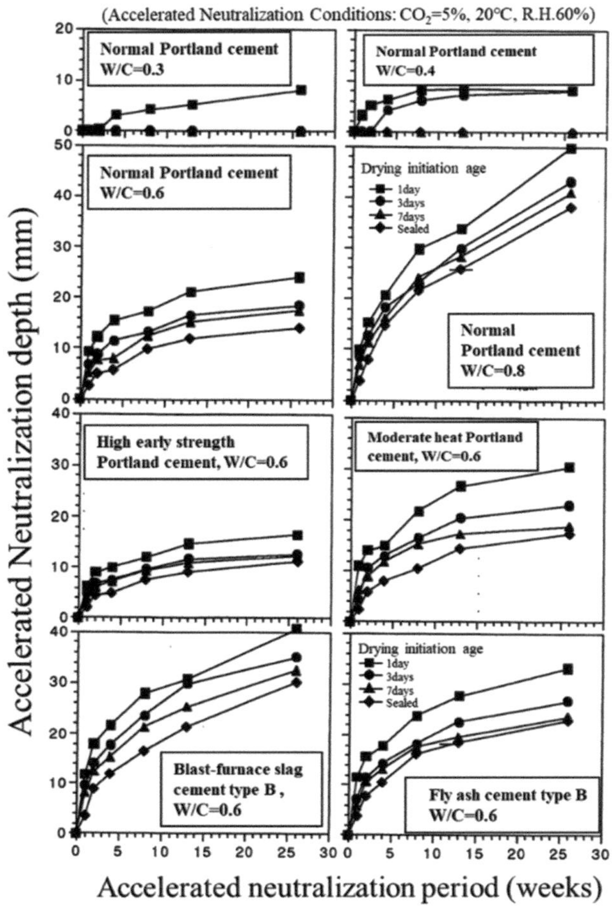

Figure 15.
Progression of neutralization of dried concrete (20°C, R.H.60%).

period, and Pf is the accelerated period (weeks) when the test is scheduled to end (= 26 (weeks)).

$$\mathrm{NF} = \mathrm{Dc} \times \mathrm{Pf}/P_{10} \tag{5}$$

It is said that there is a strong correlation between pore structure and neutralization [20]. The neutralization index obtained here should be closely related to the pore structure of surface layer concrete of 0–1 cm.

Figure 16 shows the relationship between the total effective pore volume and the neutralization index [3]. The black plots are the results for concrete using ordinary Portland cement, and the curves represent the approximate curves when the results are approximated by the power formula. For ordinary Portland cement, neutralization resistance could be evaluated by the total effective pore volume regardless of the water-cement ratio or the drying initiation age.

When other cements are used, although there is a correlation between the pore volume and the neutralization index for each cement used, the relationship between the pore volume and the neutralization index varies depending on the cement used. This is because the alkalinity after hardening differs depending on the type of cement used.

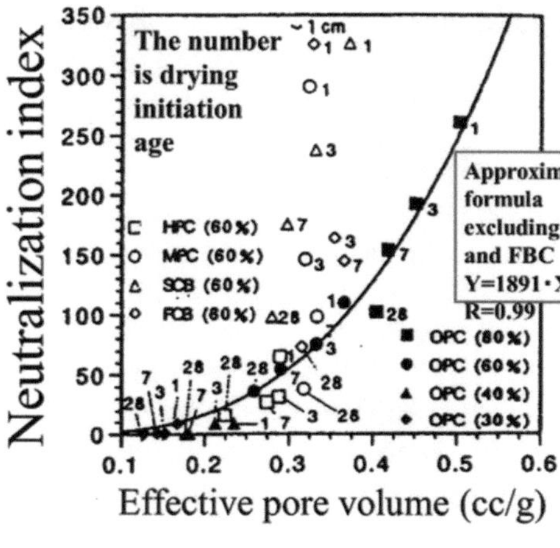

Figure 16.
Relationship between effective pore volume of 0–1 cm layer and ease of neutralization.

4.3 Isothermal moisture adsorption and desorption characteristics

Isothermal moisture adsorption and desorption characteristics is related to the moisture control property. **Figure 17** shows the relationship between the relative humidity of the surface 0–1 cm of dried concrete and the equilibrium moisture content per unit of cement paste [21].

Looking at differences in the water-cement ratio for concrete using ordinary Portland cement, for concrete with a water-cement ratio of 0.3, the difference in pore size distribution with a radius of 56 nm or less due to differences in the drying initiation age is small, and so the difference in the amount of isothermal adsorption and desorption is also small. As the water-cement ratio increases, differences are seen in the isothermal adsorption and desorption properties, corresponding to the

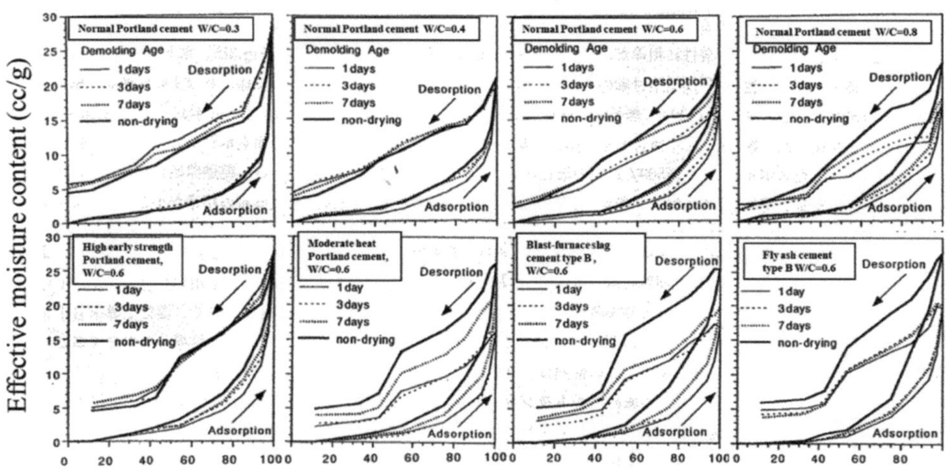

Figure 17.
Isothermal moisture absorption and desorption.

difference in pore size distribution with a radius of 56 nm or less due to differences in the drying initiation age. The later the drying initiation age is and the greater the volume of pores with a radius of 56 nm or less, the greater the content of isothermal adsorption and desorption.

Next, looking at the effect of the cement used on the relationship between relative humidity and equilibrium moisture content, differences can be seen in the content of isothermal adsorption and desorption that correspond to differences in the distribution of pore sizes with a radius of 56 nm or less due to differences in the drying initiation age. Compared to ordinary Portland cement, when moderate heat Portland cement, blast-furnace cement type B, and fly ash cement type B are used, differences in the drying initiation age have a greater effect on the content of isothermal adsorption and desorption. On the other hand, for high early-strength Portland cement, which hydrates quickly, the difference in pore structure is small, so the effect of the drying initiation age on the content of isothermal adsorption and desorption is small.

The isothermal adsorption amount of concrete can be explained by the capillary condensation theory based on its pore structure [22]. The isothermal moisture absorption and desorption of concrete is greatly affected by the type of cement used and the water-cement ratio, as well as by the inhibition of hydration caused by drying. In general, the isothermal moisture absorption and desorption amount is large in concrete that is not inhibited by drying and has many pores with a radius of 18 nm or less formed by hydration. Therefore, the isothermal moisture absorption and desorption amount is large in concrete with a high water-cement ratio that has been sufficiently wet cured.

4.4 Air permeability of surface concrete

Air permeability allows the penetration of external deterioration factors easily and is one of the basic properties of concrete.

Figure 18 shows the relationship between the drying initiation age and the air permeability speed tested by the NDIS 3436 "Concrete Surface Air Permeability Test Method" of the Japanese Society for Non-destructive Inspection [23]. The NDIS 3436 has been developed and improved from the method proposed by J.W. Figg [24]. The air permeation speed is calculated using Eq. (6) in the figure.

The air permeability speed increases as the water-cement ratio increases and as the drying initiation age is earlier.

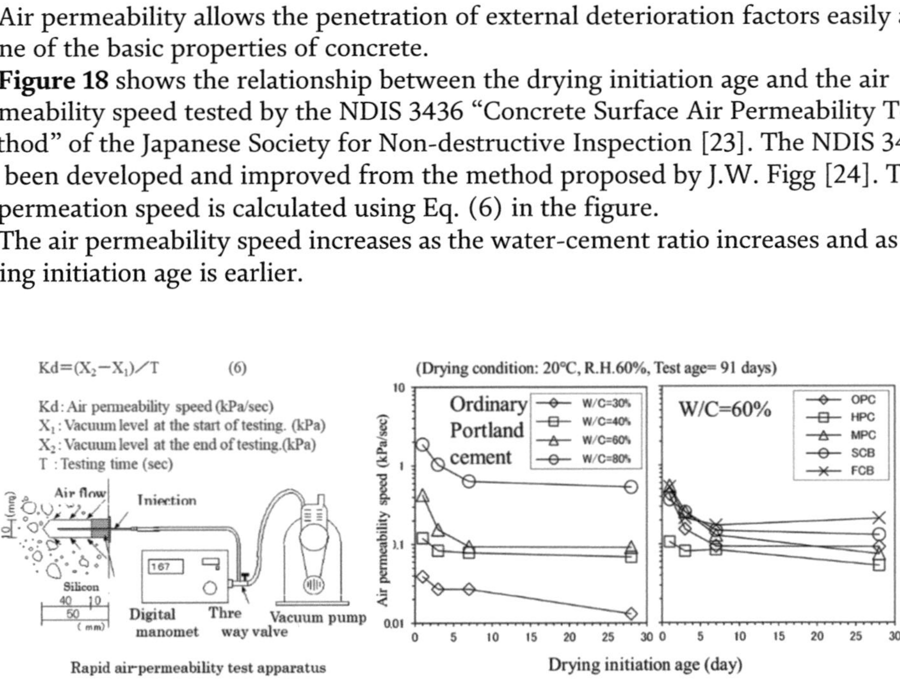

Figure 18.
Air permeability of surface concrete.

In terms of the type of cement used, the air permeability of concrete used high early-strength Portland cement, which hydrates quickly, has a few differences depending on the drying initiation age. On the other hand, the air permeability of concrete used moderate heat Portland cement, blast-furnace cement type B, and fly ash cement type B, which hydrates slowly, varies greatly depending on the drying initiation age, and increases when the concrete is dried from an early age after casting.

These results tend to coincide with the results of the double-chamber surface concrete air permeability test conducted by Torrent et al. [25], which uses a different method from Figg's method.

4.5 Semi-permeability

Long ago, Hansen attributed the mechanism of expansion destruction due to alkali-aggregate reaction to osmotic pressure [26]. McConnell et al. treated mortar as a membrane and confirmed the semi-permeability of hardened mortar [27].

Warlow et al., followed by Giinterll et al., attributed the mechanism of floor blistering to osmotic pressure due to the semi-permeability of hardened cement paste and mortar [28, 29]. **Figure 19** shows the effect of the water-cement ratio and the age at the start of the test on water movement due to the semi-permeability of the cement paste, obtained using the device shown in **Figure 20** [30]. The larger the water-cement ratio and the earlier the test is started after casting, the greater the volume of water movement due to the coarse pore structure.

When examining the water movement speed, as shown in **Figure 21**, there is a high correlation with the sum of the pore volumes of 2800-90 nm and 16-3 nm in radius. It is presumed that the 2800-90 nm in radius pores are involved in permeability, while the 16-3 nm in radius pores act to block the passage of Na^+ and Cl^-.

5. Peeling and blistering of finishing materials and quality of surface concrete

The swelling of waterproofing and blistering of floor coatings as shown in **Figure 22**, is a major construction problem. Until now, in order to prevent swelling and blistering, it was thought that "sufficient drying" of substrate concretes was the most important, and there was a tendency to welcome drying of concrete slabs immediately after casting.

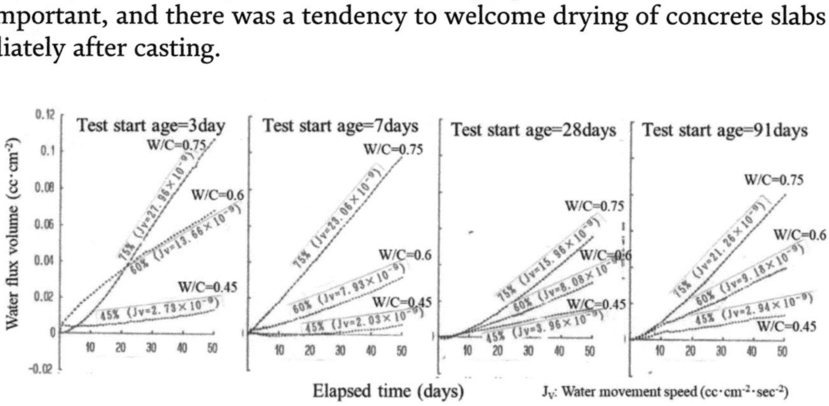

Figure 19.
Water movement through semi-permeable cement paste.

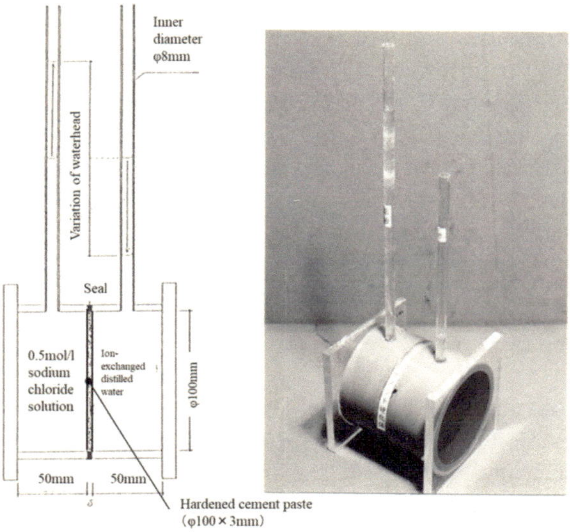

Figure 20.
Apparatus of measuring water flux.

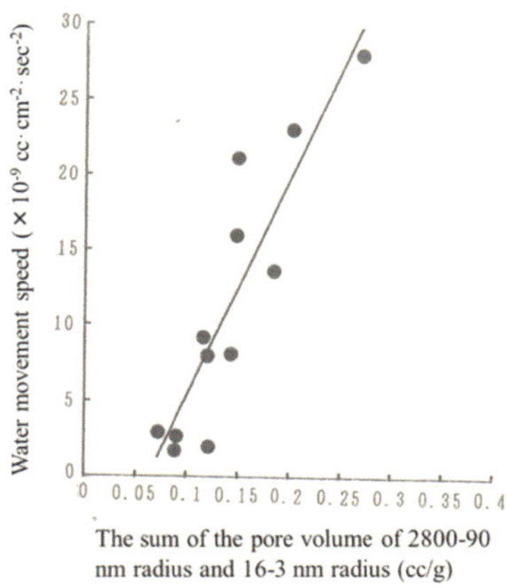

Figure 21.
Pores responsible for semi-permeability (NaCl 0.5 mol/l).

a) Swelling of waterproofing b) Blistering of floor coating

Figure 22.
Swelling and blistering of finishing materials.

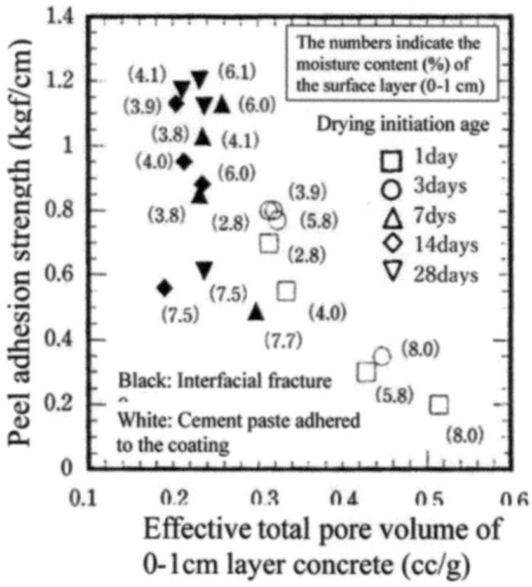

Figure 23.
Surface concrete condition and blistering rate of the painted floor layer.

As shown in **Figure 23**, when the relationship between peel adhesion strength of the coated floor and the total effective pore volume and moisture content of the surface layer 0–1 cm concrete was examined by variance analysis, firstly, it is important that the moisture content is around 6% or less [31]. Secondly, when the moisture content is below 6%, the peel adhesion strength depends on the total effective pore volume of the surface layer. From the viewpoint of preventing the blistering of finishing, it is important to reduce the water-cement ratio and to perform initial wet curing in order to densify the surface layer concrete structure.

From the standpoint of specifications, when considering the quality of the base concrete and the blistering of the coated floor, it is effective to reduce the water-cement ratio, ensure moist curing immediately after casting, and then allow a long drying period, as shown in **Figure 24** [32].

6. Conclusion

During the construction process, reinforced concrete structures are usually stripped from their forms in the early concrete age, while the cement is still hydrating.

This chapter treats concrete not as a mass, but from the surface to the interior at a resolution of approximately 1 cm, and presents the effect of early drying on the required performance of reinforced concrete structures.

When concrete dries, the moisture content immediately decreases from the surface layer, and effect until about 5 cm in depth. This phenomenon halts the densification of the pore structure, and the coarse pore structure remains for a long period of time.

Correspondingly, the strength of the surface concrete becomes smaller than that of the interior. This decrease in strength can reach more than 15 N/mm^2.

Furthermore, since the surface layer concrete is a concrete cover for reinforced concrete, its quality is important, but because it has a coarse pore structure, it allows external deterioration factors to easily penetrate, accelerating the progression of all types of deterioration.

Furthermore, a decrease in the surface strength of concrete can lead to peeling and blistering of finishing materials applied to the substrate concrete.

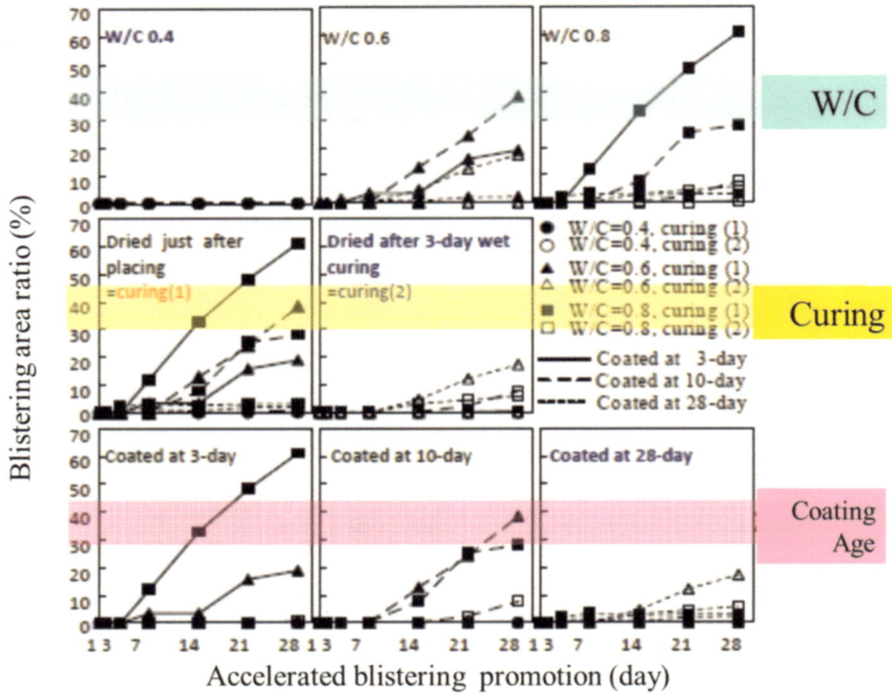

Figure 24.
Surface concrete condition and blistering rate of the painted floor layer.

It is true that the quality of concrete is primarily determined by the water-cement ratio, but this is not absolute and it is merely aimed at the potential quality of the concrete deeper than 5 cm. The quality of surface concrete can sometimes correspond to the quality of concrete with a water-cement ratio 20% higher.

It is hoped that concrete design will be considered in the future based on these results. In the field of non-destructive testing of structures, most test methods are surface-approach methods, which must have limitations in evaluating internal quality [33]. In the field of non-destructive testing, we would like to see a review of the system based on the premise that "surface quality differs from internal quality" [3].

Author details

Noboru Yuasa
Nihon University, Tokyo, Japan

*Address all correspondence to: yuasa.noboru@nihon-u.ac.jp

References

[1] Gilkey HJ. The effect of varied curing conditions upon the compressive strength and mortars and concrete. Journal Proceedings. 1926;**22**:395-431

[2] Yuasa N, Kasai Y, Matsui I. Inhomogeneous Distribution of Moisture Content and Porosity in Concrete, Creating with Concrete, Concrete Durability and Repair Technology. Institution of Civil Engineers; 1999. pp. 93-101

[3] Yuasa N. State of the art of non-destructive testing methods for concrete structures in Japan. Journal of the College of Industrial Technology, Nihon University. 2024;**57**(1):11-31. Available at: https://www.cit.nihon-u.ac.jp/laboratorydata/kenkyu/publication/journal_a/a57.1.2.pdf

[4] Yoshino T, Kamada E, Tabata M, Yanagi T. Estimating of concrete strength based on pore structure consideration: Part I, correlation between compressive strength and pore structure. Transactions of Architectural Institute of Japan. 1982; **312**:9-17 (in Japanese)

[5] Sakata K, Kuramoto O. A study on the water diffusion and shrinkage in concrete by drying. Japanese Journal of JSCE. 1981;**316**:145-152. DOI: 10.2208/jscej1969.1981.316_145 (in Japanese)

[6] Chen X, Wu S. Influence of water-to-cement ratio and curing period on pore structure of cement mortar. Construction and Building Materials. 2013;**38**:804-812

[7] Shimomura T, Maekawa K. Analysis of the drying shrinkage behavior of concrete using a micromechanical model based on the micropore structure of concrete. Concrete Library of JSCE. 1996;**27**:122-144

[8] Kumara R, Bhattacharjeeb B. Porosity, pore size distribution and in situ strength of concrete. Cement and Concrete Research. 2003;**33**:155-164

[9] Yuasa N, Kasai Y, Matsui I, Okawara O. Effect of drying condition on the hydration, porosity and strength of a small cement paste specimen. AIJ Journal of Structural and Construction Engineering. 1998;**505**:15-21 (in Japanese)

[10] Yuasa N, Kasai Y, Matsui I. Inhomogeneous distribution of compressive strength from the surface layer to internal parts of concrete in structures. In: CANMET/ACI/JCI, Fifth International Conference on Durability of Concrete. 2000. pp. 269-281

[11] Yuasa N. Non- destructive testing methods for water content of concrete: State-of-the-art in Japan. Seventh US-Japan NDT Symposium. 2022;**8**

[12] Yuasa N, Kasai Y, Matsui I, Kamada E. Freeze-Thaw Resistance of Concretes Subjected to Drying at Early Ages, Thomas Telford Limited Challenges of Concrete Construction, Concrete Technology Unit, University of Dundee, Concrete for Extreme Conditions. 2002. pp. 437-446

[13] Helmuth RA. Capillary Size Restrictions on Ice Formation in Hardened Portland Cement Pastes, Fourth International Symposium on the Chemistry of Cement, Washington; 1960

[14] Kamada E. Frost damage of concrete and pore structure of hardened cement paste. Proceedings of the Japan Concrete Institute. 1988;**10**(1):51-60 (in Japanese)

[15] Kamada E, Senbu O, Tabata M, Tanaka H. Statistical investigation

concerning the effects of pore structure on the frost resistance of concrete. Journal of Structure and Construction Engineering, Architectural Institute of Japan. 1996;**487**:1-8 (in Japanese)

[16] Kamada E, Koh E, Takahashi A. Freeze-thaw resistance and porosity of hardened cement paste used every types of cement. In: Proceeding of the 35th Meeting of Hokkaido Branch of Architectural Institute of Japan. 1971. pp. 7-10 (in Japanese)

[17] Trinker BD. et al. Methods of Structure Resistance in Estimating the Life of Heavy Concretes, Primary Report of RILEM Symposium, Durability of Concrete III; 1969

[18] Kokubu K. Fundamental studies on freeze-thaw durability of expansive concrete, Japan Society of Civil Engineers. Proceedings of the Japan Society of Civil Engineers. 1983;**334**: 145-154 (in Japanese)

[19] Sumiyoshi H, Kuboyama K, Imahashi T, Shioya M. Effect of steam curing on micro structure and other properties of concrete, the cement Association of Japan. Review of the 25th General Meeting – Technical Session. 1981;**35**:151-153 (in Japanese)

[20] Gonen T, Yazicioglu S. The influence of compaction pores on sorptivity and carbonation of concrete. Construction and Building Materials. 2007;**21**: 1040-1045

[21] Yuasa N, Kasai Y, Matsui I. Sorption isotherm and moisture diffusivity of surface layer concrete. Cement Science and Concrete Technology. 1998;**52**: 1042-1049 (in Japanese)

[22] Espinosa RM, Franke L. Influence of the age and drying process on pore structure and sorption isotherms of hardened cement paste. Cement and Concrete Research. 2006;**36**:1969-1984

[23] Nonaka A, Yuasa N. Rapid testing method for air-permeability of concrete in structure using borehole. In: 4th International Symposium on Life-Cycle Civil Engineering. CRC Press; 2014. pp. 1917-1922

[24] Figg JW. Methods of measuring the air and water permeability of concrete. Magazine of Concrete Research. 1973; **25**(85):213-218

[25] Torrent R, J, Neves RD, Imamoto K. Concrete Permeability and Durability Performance Form Theory to Field Applications. CRC Press, Modern Concrete Technology 21; 2022

[26] Hanson WC. Studies relating to the mechanics by which the alkali-aggregate reaction produces expansion in concrete. Journal of the American Concrete Institute. 1944;**40**:213-228

[27] McConnell D et al. Cement-aggregate reaction in concrete. Journal of the American Concrete Institute. 1947; **44**:93-128

[28] Warlow WJ, Pye PW. Osmosis as a cause of blistering of in-situ resin flooring on wet concrete. Magazine of Concrete Research. 1978;**30**

[29] Gunter M. Influence of physical and chemical interactions between a concrete substrate and organic surface coatings on bond strength. Durability of Non-Metallic Inorganic Building Materials. 1988

[30] Yuasa N, Michio K, Tanaka K, Hashida H. Water flux by semi-permeability of hardened cement paste – Effect of porosity of hardened cement paste. Architectural Institute of Japan,

Summaries Technical Papers of Annual Meeting. 1990:227-228 (in Japanese)

[31] Yuasa N, Kasai Y, Matsui I, Henmi Y, Sato H. Effect of moisture content and pore structure of substrate concrete on adhesive strength of epoxy coating. In: The Second East Asia Symposium on Polymers in Concrete. CRC Press; 1997. pp. 401-408

[32] Yuasa N, Sasaki T, Matsui I, Kasai Y. Effect of porosity and moisture content of concrete slab on osmotic blister of polymer. In: 6th International Colloquium Industrial Floors'07. Vol. 1. 2007. pp. 269-274

[33] Yuasa N. Comparison between rebound numbers using various rebound hammers and proposal of requesting method for relationship between rebound numbers and compressive strength. In: TAE 4th International Symposium on Life-Cycle Civil Engineering. 2014. pp. 1163-1169